AFRICAN GREENHOUSE GAS EMISSION INVENTORIES AND MITIGATION OPTIONS: FORESTRY, LAND-USE CHANGE, AND AGRICULTURE

Johannesburg, South Africa
29 May - 2 June 1995

Edited by

John F. Fitzgerald
U.S. Country Studies Program, Washington, D.C., USA
Barbara V. Braatz
ICF Incorporated, Washington, D.C., USA
Sandra Brown
U.S. Environmental Protection Agency, Corvallis, Oregon, USA
Augustine O. Isichei
Obafemi Awolowo University, Ile-Ife, Nigeria
Eric O. Odada
University of Nairobi, Nairobi, Kenia
Robert J. Scholes
Division of Forest Science and Technology, CSIR, South Africa

Sponsored by:

United States Country Studies Program
United Nations Environmental Programme

Reprinted from *Environmental Monitoring and Assessment*, Volume 38, Nos. 2-3, 1995

Kluwer Academic Publishers
Dordrecht / Boston / London

A C.I.P. Catalogue record for this book is available from the Library of Congress

ISBN-13: 978-94-010-7232-8 e-ISBN-13: 978-94-009-1637-1
DOI: 10.1007/978-94-009-1637-1

Published by Kluwer Academic Publishers,
P.O. Box 17, 3300 AA Dordrecht, The Netherlands.

Kluwer Academic Publishers incorporates
the publishing programmes of
D. Reidel, Martinus Nijhoff, Dr W. Junk and MTP Press.

Sold and distributed in the U.S.A. and Canada
by Kluwer Academic Publishers,
101 Philip Drive, Norwell, MA 02061, U.S.A.

In all other countries, sold and distributed
by Kluwer Academic Publishers Group,
P.O. Box 322, 3300 AH Dordrecht, The Netherlands.

Cover design: Photograph of Earth from Apollo 17
Courtesy of U.S. National Aeronautics and Space Administration

Printed on acid-free paper

TABLE OF CONTENTS

FOREWORD

The ultimate objective of the United Nations Framework Convention on Climate Change (FCCC) is the stabilization of atmospheric concentrations of greenhouse gases, but a number of important problems remain to be solved before a target stabilization level can be quantitatively specified. What *can* be calculated with some confidence is that current emissions from human activities are likely to lead to increasing atmospheric concentrations of these gases over coming decades or even centuries. Whatever route is eventually chosen to meet the Convention's ultimate objective, systematic quantification of present day anthropogenic emissions, using comparable methods, is a necessary basis for international agreements on future emissions.

Once compiled, a national greenhouse gas inventory serves as a valuable tool within the context of the FCCC. It allows each country to place its own emissions within the larger picture of global emissions, and provides a baseline against which its own future emissions can be compared. It also provides a basis for the formulation of a national greenhouse gas mitigation policy. Furthermore, experience shows—and the studies reported here are no exception—that compilation of the inventory brings additional benefits, including improved national statistics, increased awareness of the issues surrounding climate change among government and industry, and improved cooperation between countries with similar patterns of greenhouse gas emissions.

The U.S. Country Studies Program and the United Nations Environment Programme/ Global Environment Facility (UNEP/GEF) Country Case Studies Project use the Intergovernmental Panel on Climate Change (IPCC) Guidelines for National Greenhouse Gas Inventories as their standard inventory methodology. The guidelines currently represent the only inventory methodology specifically accepted by the parties to the FCCC. Developed over three years under a program coordinated jointly by IPCC, the Organisation for Economic Co-operation and Development (OECD), and the International Energy Agency (IEA), the guidelines provide an internationally accepted methodology for quantifying emissions of carbon dioxide, methane, and nitrous oxide. Methods and default data contained in the guidelines are based on wide international consultation and on the best information available in the scientific and technical literature. Nevertheless, the development of the guidelines must be seen as an ongoing process, and the studies reported in this workshop represent searching tests of the practical applicability of the guidelines as well as important sources of guidance for future improvements.

Under the IPCC guidelines, the basic approach for calculating emissions of a particular gas from a particular sector or sub-sector is simple in concept:

$$\text{Emissions} = \text{Activity Level} \times \text{Emission Factor}$$

However, determining activity levels and emission factors can present difficulties. A particular feature of greenhouse gas emissions in Africa is the high proportion that comes from agriculture, land-use change, and other area sources, precisely the categories that present the greatest practical difficulties for inventory calculations. Many of the activity data—for example, area of savanna burned per year—either do not exist or must be derived from related statistics, or even from anecdotal evidence. Experimentally based emission factors specific to a particular region or system are scarce, so those factors that *are* known are often applied in places that may be beyond their region of validity. Compilation of a national greenhouse gas inventory, nowhere easy, thus presents particular problems for countries in Africa. Papers presented in this volume demonstrate a considerable degree of tenacity in extracting activity data from disparate sources, and also report important progress in the exploitation of remote sensing and in the improved quantification of emission factors.

Finally, it is very important to acknowledge the coordinating and facilitating role played by the various country study programs related to the Climate Convention, including the U.S. Country Studies Program and the UNEP/GEF Country Case Studies Project. Through input of resources and through collaborative efforts, these programs supported the original development of the IPCC guidelines, helped to achieve their wide international acceptance, and contributed to their final publication. In addition, country case study programs that have used the IPCC methodology have also been instrumental in delivering the IPCC manuals, and providing training in their use, to a large number of developing countries and countries with economies in transition. The substantial progress in the compilation of national greenhouse gas inventories that was evident in the Johannesburg workshop represents some of the rewards of these extensive international efforts.

Bruce A. Callander
Intergovernmental Panel on Climate Change
Meteorological Office, Hadley Climate Centre
London Road, Bracknell
RG12 2SY, United Kingdom

PREFACE

In an effort to address greenhouse gas emission inventory development and emissions mitigation in an African context, the U.S. Country Studies Program and the United Nations Environment Programme/Global Environment Facility Country Case Studies Project cosponsored a workshop held near Johannesburg, South Africa from May 29 to June 2, 1994. The workshop was attended by sixty-four scientists and policy experts from eighteen African and five non-African nations. Papers were presented on methodological and scientific aspects of emission estimation and mitigation assessment, on the results of national inventory and mitigation analyses, and on resource and programmatic issues. In addition, three independent working groups were formed to discuss and make recommendations on (1) Forestry and Land-Use Change, (2) Vegetation Burning, and (3) Livestock.

This volume is composed of thirteen papers and two short communications in four sections. Part I, the Workshop Summary Statement, presents the main conclusions of discussions in plenary and in working groups, followed by the three working group summaries. Part II, International Context, presents two papers that describe important scientific aspects of the United Nations Framework Convention on Climate Change and outline a number of significant international efforts to address climate change. Papers in Part III, Greenhouse Gas Emission Inventories and Mitigation Options: Methodological Issues, and Part IV, Greenhouse Gas Emission Inventories: National Assessments, discuss methodological and scientific issues associated with inventory assessment and report the experiences of some workshop participants in conducting greenhouse gas emission inventories in Africa.

It is hoped that this volume, and the workshop from which it was derived, will stimulate future interest in the climate change issues that are relevant to Africa and encourage further African involvement in developing greenhouse gas emission inventories and mitigation assessments in support of the United Nations Framework Convention on Climate Change.

John F. Fitzgerald
Barbara V. Braatz
Sandra Brown
Augustine O. Isichei
Eric O. Odada
Robert J. Scholes

ACKNOWLEDGMENTS

We wish to acknowledge with sincere thanks all of the workshop participants, whose hard work and enthusiasm made this publication possible. Their ideas, knowledge, and insights provide the substance of this book.

We are grateful to the U.S. Country Studies Program (U.S. CSP) and the United Nations Environment Programme/Global Environment Facility Country Case Studies Project (UNEP) for their financial support of the workshop. From the funding organizations, we would particularly like to thank Michael Short (UNEP) for his enthusiastic participation in the organization of this workshop. We also wish to acknowledge Robert Dixon of the U.S. CSP for his helpful comments and direction throughout the workshop process. The information in this document has been subjected to technical peer review, but does not necessarily reflect the official views of any governmental or intergovernmental body.

For their untiring effort in organizing and coordinating the workshop, we would like to thank Rebecca Holmes and Susan Barvenik of ICF Incorporated, and Dawn Middleton, Marna ven der Merwe, Cathy Bailey, Shane Abbott, and Minoda Mathura of Forestek, CSIR. We thank William Driscoll of ICF Incorporated for his invaluable editorial assistance and for coordinating the peer review process. The front cover design was developed by John Paul McCarty (ICF Incorporated). Final formatting and production of this volume were completed with support from Kimmie Shreve-Hilten, Joan Myers, and the Publications Department of Enviro-Management & Research, Incorporated and Emily Barnett, Ellie Stewart, and Jim Woldahl (ICF Incorporated). We also thank Mariette de Jong of Kluwer Academic Publishers for her publication assistance.

John F. Fitzgerald
Barbara V. Braatz
Sandra Brown
Augustine O. Isichei
Eric O. Odada
Robert J. Scholes

PART I

WORKSHOP SUMMARY STATEMENT

AFRICAN GREENHOUSE GAS EMISSION INVENTORIES AND MITIGATION OPTIONS: FORESTRY, LAND-USE CHANGE, AND AGRICULTURE

B.V. BRAATZ[1], S. BROWN[2], A.O. ISICHEI[3], E.O. ODADA[4], R.J. SCHOLES[5], Y. SOKONA[6], P. DRICHI[7], G. GASTON[2], R. DELMAS[8], R. HOLMES[1], S. AMOUS[9], R.S. MUYUNGI[10], A. DE JODE[11], AND M. GIBBS[12]

[1]ICF Incorporated, 1850 K St., N.W., Suite 1000, Washington, D.C. 20006, USA. [2]USEPA, Environmental Research Laboratory, 200 SW 35th St., Corvallis, OR 97333, USA. [3]Botany Department, Obafemi Awolowo University, Ile-Ife, Nigeria. [4]START/ENRICH Africa Liaison, Ghana Academy of Sciences, Agustino Neto Road, P.O. Box M32, Accra, Ghana. [5]Forestek, CSIR, P.O. Box 395, Pretoria 0001, South Africa. [6]ENDA - Energy, PB 3370, 54 rue Carnot, Dakar, Senegal. [7]National Biomass Study, Department of Forestry, P.O. Box 1613, Kampala, Uganda. [8]Laboratoire d'Aerologie, U.R.A./C.N.R.S. 354, Université Paul Sabatier, 118 route de Narbonne -F - 31062, Toulouse Cedex, France. [9]APEX-EBBD Consulting Co., 7 rue Malta Sghira, 1000 RP, Tunis, Tunisia. [10] The Centre for Energy, Environment, Science, and Technology (CEEST), 1372 Karume Road, Oysterbay, PO Box 5511, Dar es Salaam, Tanzania. [11]Appropriate Technology International, 67 East Ave., Oxford, United Kingdom, OX4 1XR. [12]ICF Incorporated, 10 Universal City Plaza, #2400, Universal City, CA 91608, USA.

1. Background and objectives

Human-induced change in the composition of the atmosphere seriously threatens the global climate. In an effort to address this threat, 161 nations signed the United Nations Framework Convention on Climate Change at the Earth Summit in Rio de Janeiro in June 1992. The ultimate objective of this international agreement is to *achieve stabilization of greenhouse gas concentrations in the atmosphere at a level that would prevent dangerous anthropogenic interference with the climate system.* As of the first session of the Conference of the Parties (March/April 1995), 128 nations had ratified the Convention. Parties to the Convention commit to numerous obligations, including commitments to: *develop national inventories of anthropogenic emissions using agreed, comparable methodologies*; and to *formulate and implement national programs containing measures to mitigate, and to facilitate adaptation to, climate change.* Several bilateral and multilateral programs, including the United States Country Studies Program (U.S. CSP) and the United Nations Environment Programme/Global Environment Facility Country Case Studies Project (UNEP), have been established to provide technical and financial assistance to developing countries and countries with economies in transition in meeting their commitments under the Convention.

From 29 May to 2 June 1995, a workshop cosponsored by U.S. CSP and UNEP was held near Johannesburg, South Africa to address emission inventory development and emissions mitigation in an African context. The workshop focused on forestry, land-use

Environmental Monitoring and Assessment **38**: 109-126,1995.

change, and agriculture because these "sectors" play a major role in generating greenhouse gas (GHG) emissions from the continent, provide promising opportunities for emission mitigation, and are a vital component of African growth and development. The primary objectives of the workshop were to:

- Promote the exchange of information based on participants' experience in preparing national inventories and assessing mitigation options
- Further the development of consensus among experts on emission estimation and mitigation methods suitable for African countries
- Identify priority needs for support in testing and using the methodologies
- Identify and discuss possible options for mitigating emissions in Africa
- Promote cooperation and coordination among participants and institutions involved in climate change studies in Africa

The workshop was attended by 64 governmental and nongovernmental scientists and policy experts from 18 African and 5 non-African nations. There were 20 presentations by specialists dealing with methodological and scientific aspects of emission estimation or mitigation, 9 presentations reporting the experiences of African countries in conducting inventories or mitigation assessments, and 6 presentations from organizations providing financial, technical, or logistic support to the inventory and mitigation assessment process. Three working groups were formed to discuss and make recommendations on (1) Forestry and Land-Use Change, (2) Vegetation Burning, and (3) Livestock.

This workshop statement presents the main conclusions of discussions in plenary and in working groups, followed by the three working group summaries. The formal publication resulting from the workshop contains this statement, as well as individual papers contributed by participants (see inside front cover for details).

2. Main conclusions

The workshop reemphasized the importance of non-point sources of GHGs (i.e., forestry, land-use change, vegetation burning, and ruminant animals, in contrast to "point sources" such as fossil energy combustion) in the emission inventories of African countries, and pointed out a number of obstacles which stand in the way of their accurate estimation. In general, the inventory methods contained in the *IPCC Guidelines for National Greenhouse Gas Inventories* (UNEP/OECD/IEA/IPCC, 1995) for non-point sources are not as well-developed and tested as those for point sources. In most cases the methods themselves are acceptable, although the scope of the methods should be expanded to allow for more detailed and comprehensive assessments and for inclusion of additional gases. More importantly, the data needed to implement the methods are either unavailable or inadequate, and the default values provided by the IPCC are too general for accurate assessments.

The workshop participants noted that the completion of national emission inventories is not, in general, a high priority in countries confronted by an urgent need to develop, but that the process of emission estimation could be complementary to development imperatives. For instance, the inventory process has highlighted the poor state of national-level statistics in many African countries. These statistics, which include items such as the distribution of agricultural practices, the state of forests, and the sizes of livestock herds, have uses far beyond the preparation of emission estimates. Participants agreed that these statistics should be routinely maintained by national governments because the same statistics that are needed to produce a GHG inventory are also necessary for effective development planning.

The analysis of mitigation options must be closely linked to emission inventory assessment. Most importantly, mitigation options must be consistent with, and complementary to, development objectives in order to be viable.

Greater coordination among the various agencies providing support for inventory and mitigation activities, as well as between these agencies and other governmental and nongovernmental entities, is needed to effectively leverage limited resources. In addition, more collaboration among countries on a regional basis should be promoted.

In particular, participants agreed that there should be more African involvement in further development and implementation of inventory and mitigation methodologies.

3. Individual working group reports

3.1. Forestry and Land-Use Change

Co-Chairs: Youba Sokona & Paul Drichi
Rapporteurs: Barbara V. Braatz & Greg Gaston

3.1.1. Emission inventory scope

Forestry and land-use change is one of the most significant sectors in Africa, both in the context of GHG emissions and as a critical part of African development and planning. This sector is a major source of GHG emissions in many African countries primarily because of rapid rates of deforestation and forest degradation from cropland establishment and timber and fuelwood collection. It also presents opportunities for emission mitigation through both reducing emissions and increasing sinks. Forests and other woody biomass stocks play a critical role in the energy sector, as fuelwood and charcoal are the principal sources of energy for many countries in Africa. This sector is also an important source of food, fodder, construction material, and medicinal plants, provides jobs and income as well as foreign exchange, and plays a critical role in biodiversity conservation.

Despite general agreement that harvest and removal activities far outpace the reestablishment of forests and other woody biomass stocks in most of Africa, explicit proof is lacking in many regions. At the same time, extrapolations that predict no forests in Africa in the near future are unlikely. A thorough GHG emission inventory can be a very useful tool for defining forest resources, rates of use, and land-use changes; and the same data that are needed to produce an accurate inventory are also necessary for effective development planning. However, at present, inventories for this sector are highly uncertain due to complexities associated with emission-producing activities, inventory methodologies, and the data needed to implement methodologies.

Overall, the sources and sinks in the forestry and land-use change sector are fairly well defined in the *IPCC Guidelines*, although two issues require further clarification. The first is "carbon uptake" and the discussion in the *Guidelines* of how this should be tracked in the calculations. The second is the connection between this sector and the energy sector. In particular, accounting for non-CO_2 emissions from woodfuel use in the energy sector and CO_2 emissions from woodfuel that contributes to forest clearing in the Forestry and Land-Use Change sector is confusing.

Although this sector is meant to cover all land-use change activities, as well as all activities that affect woody biomass stocks, there are several activities that are particularly important to Africa, but that are not discussed in detail in the *Guidelines*. In particular, trees on small farms can be a sink of carbon when they are growing and a source of carbon when they are harvested for use as fuelwood. Conversion of non-forest natural vegetation (e.g., rangelands, savannas) to croplands is another potentially important source of carbon. The *Guidelines* would be improved by more detailed treatment of these activities. At the same time, because of difficulties associated with collecting data (particularly historical data), there will always be a trade-off between greater detail in emission assessments and the available resources with which to collect and analyze data. The relative magnitudes of sources and sinks should be considered when planning and implementing how to allocate resources to an inventory assessment.

Changes in soil organic carbon and belowground biomass as a result of land-use change contribute significantly to carbon flux. At present, soil carbon is only partially included in the IPCC methods, while belowground biomass is not included at all. These components of overall carbon stocks should be considered in further improvements in the *Guidelines*, although data limitations will likely continue to impede full implementation.

3.1.2. Emission inventory data

Completion of a comprehensive inventory for forestry and land-use change requires both a wide variety and a large amount of data, both for the inventory year and for historical years. Specific data required to complete an inventory for this sector include: statistics on areas and rates of change for different land-use types, biomass densities, wood consumption statistics, and soil carbon and nitrogen data. With few exceptions within the African continent, many of the necessary data are lacking, or, if available, are of uncertain quality. Simply identifying the agency responsible for collecting and storing relevant data may be difficult. While public institutions should, in theory, be willing to share data, this is often not the case. Data from the private sector can be particularly difficult to acquire. Also, data bases within individual countries have been found to be inconsistent, and the agencies or groups responsible for collecting these data often do not assess the reasons for these inconsistencies. Although national surveys, monitoring, and data base processing and development are clearly needed, time and resource constraints will always limit improvements. At the same time, the utility of emission inventory data to national development planning must be emphasized. Several countries have already performed thorough surveys for national emission assessments. Existing national data bases that have been developed for other purposes can be adapted for use in inventory assessments.

International data bases are often inaccurate and of poorer quality than those available in-country. The FAO data sets, some of which are used as default values in the *IPCC Guidelines*, have a number of problems. For example, the FAO biomass density data contain estimates of forest (tree) biomass only, rather than whole ecosystem biomass (e.g., understory vegetation, litter, etc.), and areas and rates of change statistics for some countries have been found to be inconsistent with national data. In addition, comprehensive and comparable soils data in Africa are lacking.

Remote sensing data are a potential source of information on rates of land-use change. However, while remote sensing data provide systematic and synoptic information suitable for assessing land-cover/land-use changes, a great variety of other data are required to produce comprehensive and accurate emission inventories.

Commercial forest inventories provide estimates of potential harvestable timber, but their use for biomass density estimation is limited. Specifically, timber inventories contain a bias towards large trees of commercial species, ignoring small trees and shrubs and noncommercial species. Therefore, use of these data to assess biomass densities results in underestimates. Similarly, commercial timber harvest statistics do not include all the harvesting that occurs in the informal sector, such as wood for pit props, fuelwood, and rural construction.

Clear definitions of forest and other land-cover types, and land-use classification systems are needed. The forest classification system adopted in the *IPCC Guidelines* is not well defined, and is too generic to be usefully applied to the wide variety of forest types across the African continent. The problem with definitions is partly addressed by the fact that countries are encouraged to use whatever classification system is most appropriate when developing emission inventories, with the stipulation that they clearly define this classification system in order for their inventory to be transparent and verifiable. Clear definitions of national classification systems will also facilitate sharing of data across national boundaries.

The applicability and clarity of the *Guidelines* would be improved with more comprehensive conversion factors, and clearer units. In particular, the *Guidelines* only provide one default value for wood density, which can vary significantly by species. In addition, units should be more clearly defined.

3.1.3. Emission inventory methodology

Four methodological issues associated with the *IPCC Guidelines* need to be addressed in this sector. First, the methods ignore international exports of forest products. In the fossil energy sector, however, exports are subtracted from production so that fuels that are produced in one country, but consumed in another, do not contribute to emissions in the producing country. A similar step should be included in the forestry calculations so that forest product exports do not contribute to the producing country's estimated emissions.

This omission of exports in the forestry and land-use change sector is likely to result in overestimates of CO_2 emissions in Africa since several African countries are major exporters of roundwood.

Second, the methods treat all biomass removed from "managed forests and other woody biomass stocks" as an emission in the inventory year, rather than explicitly tracking the decay of wood products. This simplification is based on the assumption that annual biomass removals are replacements for biomass stocks that have been discarded and are decaying in the inventory year. Although data limitations make it difficult to track annual fluxes from the decay of forest products, the realism of this methodological simplification should be assessed when a country undertakes an inventory assessment. Specifically, countries should assess (either qualitatively or quantitatively, depending on data availability) whether or not the sizes of forest product pools are changing significantly over time, and if so how rapidly. Data on population growth and materials substitution can be used to make this assessment.

The *Guidelines* employ a fairly simplistic compartmentalization of forestry and land-use change activities. The methods were designed to be simple to enable implementation with limited data, and the methods are essentially a framework within which countries are encouraged to develop as detailed emission estimates as available data and scientific understanding allow. However, the generic activities upon which the methods are based are not realistic, particularly in Africa where land uses are often complex and multidimensional.

There are many land-use practices in Africa that affect biomass and soil carbon stocks without an explicit land-use change. These activities affect land cover and could have a significant effect on GHG emissions, and therefore should be addressed in the inventory methods. For example, an increase in intensity of grazing (i.e., an increase in the number of cattle on a particular area) may result in reduced grass cover and increased soil erosion, and thereby enhance soil carbon emissions without an explicit land-use change.

3.1.4. Mitigation options

Forestry and land-use change provide opportunities for mitigation of emissions, but in order for any mitigation option to be viable, it must be linked to a country's overall development plan. For example, issues such as land tenure, agricultural productivity, resource development, and trade must be addressed when designing mitigation strategies. In fact, if mitigation options are presented to government policymakers solely as options to reduce GHG emissions, such options would never be implemented. Instead, analysts must determine which mitigation options are consistent with, and complementary to, national development plans, and focus on those.

Linkages between inventory and mitigation methodologies, as well as between the two processes of assessment, need to be strengthened. Inventory and mitigation methodologies have, in general, been developed separately and therefore there are inconsistencies between them. Also, since an inventory sets the baseline for a mitigation assessment, the process of assessing mitigation options must build upon the inventory. Ideally, the individuals who compile a national inventory would also be involved in the national mitigation analysis to ensure continuity and consistency.

In order to adequately evaluate mitigation options in this sector, including reforestation, afforestation, and agroforestry, accurate and comprehensive cost data are needed. Documentation of past experiences with similar projects, particularly in the Sahel, are a valuable source of information with which to assess the viability of various options.

3.1.5. Programmatic issues

Country studies programs have contributed significantly to capacity building in African countries, but there are several areas for improvement–for the donors as well as the recipient countries. Donors often do not communicate effectively with one another, so that linkages between programs do not exist. As a result of this lack of communication, efforts may be duplicated. At the same time, recipient countries have a responsibility to inform donors of previous and ongoing work in this area. As they design and implement projects, donors and recipient countries must address the need for a multidisciplinary approach. In addition, both donors and recipients need to recognize the value of local expertise, and build on this expertise in order to strengthen and broaden local capacity.

There are also managerial and organizational problems associated with project implementation on the part of recipient countries, including the selection of appropriate institutions and individuals for project implementation and follow-up. Other problems include limited access to data, bureaucratic bottlenecks in the dissemination of information and results, and lack of managerial commitment to projects.

Resources for these types of country study projects are often insufficient. More resources are needed to address climate change issues, especially in order to maintain the momentum developed during initial assessments.

3.1.6. Conclusions

- *Scope.* Forestry and land-use change is an important source of GHG emissions in Africa, provides promising opportunities for emission mitigation, and is a vital component of African economic growth and development.
- *Methods and Data.* Reliable data with which to accurately assess GHG emissions and uptake due to activities in this sector are lacking. Improvements in methodology will not address this problem. Improving the quantity, quality, and accessibility of data must be a priority.

Numerous collateral benefits can accrue to a country from assembling the data necessary to conduct an accurate inventory of GHG emissions and sinks. Reliable and recent information on land-use change and forestry can be a great benefit to development activities. Remote sensing data are a potential source of information on rates of change in land cover and land use.

- *Mitigation Options.* Mitigation options within the forestry and land-use change sector must be consistent with, and complementary to, national development plans in order to be viable.

- *Programmatic Issues.* Country studies must emphasize "team building" both to effectively leverage existing resources and to develop in-country capacity. These studies should be seen as an opportunity to develop and utilize interdisciplinary teams.

There is a serious lack of regional cooperation in climate change studies that has been a barrier to effective use of local expertise in neighboring countries. This problem results from a lack of recognition on the part of the donors as well as recipient countries of the potential benefits of this exchange and sharing of expertise. Cooperation among countries is a vital component of regional capacity building. Therefore, such barriers must be removed in order to effectively develop climate change studies in general, and emission inventories in particular.

There needs to be more African involvement with the development of the *IPCC Guidelines*. The special characteristics of Africa must be addressed by Africans.

3.2. Vegetation Burning

Co-Chairs: Augustine O. Isichei & Robert A. Delmas
Rapporteurs: Rebecca Holmes & Samir Amous

3.2.1. Emission inventory scope

Emissions of GHGs from vegetation burning are very important in Africa; in many countries, they may be the dominant source of emissions. Net pyrogenic emissions include CO, CH_4, NO_x, N_2O, non-methane hydrocarbons (NMHCs), and particles. (Carbon dioxide is currently not treated as a net emission because it is reabsorbed by regrowth of the vegetation.) Even in attempting to measure emissions from other sources, it is very difficult to avoid picking up the pervasive emissions from fires. Emissions in Africa are greatest from the moist, infertile broad-leafed savanna regions which lie between the tropical rainforests and the arid savannas and subtropical deserts.

There is no clear picture of whether savanna burning is increasing or decreasing as an emission source. Over the long term (approximately 50 years), savanna burning is likely to decrease as a result of progressive change in agricultural practices. Over the short term (one to two decades), savanna burning is likely to continue increasing due to population pressure and land-cover conversion. Without active intervention, major changes in the frequency of savanna fires are not expected. Slash-and-burn practices are likely to peak in the short term and then decrease significantly in the medium term (approximately 30 years) due to shortage of land, a result of high population growth rates.

Vegetation burning as defined in the IPCC methods should be broadened to include not just savanna, but any vegetation category that is frequently burned and fire-adapted (e.g., grasslands and thickets).

Some emissions, such as NMHCs and aerosols (i.e., airborne particles) are not currently included in the *IPCC Guidelines* for vegetation burning. Unburned savanna and forest vegetation is a significant biogenic source of NMHCs and NO_x, both of which are ozone precursors. Burning may increase emissions of these gases and concentrate them into a plume with the other gases necessary for ozone formation. The IPCC methods could be improved by including aerosol emissions. Aerosol emissions from fires are likely to be significant both as antidotes to greenhouse radiant forcing (directly and indirectly through cloud nucleation) and as sites of ozone formation and/or depletion. Some data on aerosol emissions are available, but additional work is needed to characterize these emissions. The IPCC should also include NMHCs in the inventory methods to be consistent with the inclusion of other ozone precursors (CO and NO_x).

3.2.2. Emission inventory methodology and data

Savanna Burning. The basic methodology presented in the *IPCC Guidelines* (i.e., estimating emissions using data on area burned, fuel load, and emission factors) is appropriate. The method as presented is unnecessarily complicated, and could be simplified to an expression such as:

$$E = B \times BFB \times EF$$

Where	E	=	emissions
	B	=	biomass (dry matter)
	BFB	=	biomass fraction burned
	EF	=	emission factor (g/kg dry fuel)

Better data are available on emission factors than on fuel load and area burned, therefore research should focus on estimating areas and characterizing fuel loads. It is important to differentiate between different savanna types in calculating emissions. Fuel load and fuel type may vary significantly between different types of savannas varying in vegetation composition and structure, soil type, and climate. Reasonable fuel load estimates can be derived using simple algorithms relating fuel load to rainfall, soil quality, vegetation type, and herbivory, provided that these algorithms can be calibrated.

In estimating areas of savanna burned, remote sensing should be used, because anecdotal ground-based data are generally overestimates. Use of low-resolution remote sensing (such as one kilometer AVHRR) is cost-effective, but calibration of the fire counts against high-resolution images (e.g. LANDSAT, SPOT, or aerial photographs) is essential.

The model of combustible biomass growth and burning emissions developed for southern Africa (Scholes, 1995) could be applied to West Africa, but the production functions and fire detection algorithms would need to be recalibrated. Together, the Southern and West African versions of the model could be used to derive estimates of area burned and fuel load for all of Sub-Saharan Africa at a resolution satisfactory for national inventories.

The *IPCC Guidelines* assume that there is no net long-term change in carbon storage associated with savanna burning. However, there is strong evidence that if savanna fires were suppressed, woody biomass would increase and there would be an increase in carbon storage in both biomass and soil. Because the frequency of fires is not known to be changing significantly and consistently at an Africa-wide scale, it is not known if there is a net storage or release of CO_2 from savannas. Some local savanna regions are clearly decreasing in carbon density due to overapplication of fire and other reasons; others are accumulating carbon due to fire exclusion. Major changes in carbon fluxes are not likely to occur in the short term, although relevant and adequate policies to reduce savanna fires could result in a large carbon sink.

The great majority of African fires are anthropogenic, and probably have been for many centuries. Therefore, all fires should be considered in calculating emissions, not only those considered "unnatural."

Slash-and-Burn Agriculture. Fuel loads for slash-and-burn agriculture vary greatly. Since the size of the individual plots being cleared is so small, remote sensing would not be a practical tool to measure their area or characterize the carbon stock. One alternative is to use remote sensing to identify areas where slash-and-burn agriculture is practiced, based on their characteristic texture, and use ground teams to estimate the fraction of the landscape in different stages of clearing and regeneration, the biomass associated with each, and the duration of the cropping and regrowth cycle. More detailed information is also required on the state of maturity of vegetation and the fallow cycle. In addition, countries should seek information from their national agricultural services, and also increase the frequency of national land-cover mapping to at least once every 10 years.

Agricultural Residues. Emissions from field burning of agricultural residue are not significant in comparison with emissions from other sources. Most agricultural residues produced in African countries are used for energy, animal feed, the pulp and paper industry, and construction. Thus, little is burned in the field, with the exception of cotton-plant residue and the pre-harvest burning of sugar cane. In general, however, there are few reliable data available on quantities of residues produced and the proportions burned in the field. The few estimates available are based on anecdote or expert opinion rather than survey data. Research on GHG emissions from this source is necessary, but is not a high priority.

Emission Factors. Several studies have been conducted in West and Southern Africa on emissions from forest clearing, savanna burning, biofuel combustion, and charcoal production. Good data are available on emissions from savanna burning, and relatively good data are available on emissions from other sources such as forest clearing. The estimated emission factors for savannas are comparable to the default values presented in the *IPCC Guidelines*. The *IPCC Guidelines* present similar factors for forests and savannas, which the newer evidence shows are significantly different; this should be corrected. In addition, the newly available data have reduced the uncertainty range in emission factors to +/- 15%, less than that presented in the *Guidelines*. Additional research on emission factors for savanna burning should not be a high priority, with the exception of certain poorly-quantified situations such as wet-season fires.

There is much greater uncertainty in estimates of area burned and fuel loads than in emission factors. In order to conduct a complete error analysis, information on uncertainty in all parameters would be required. The IPCC recommends analysis of uncertainty, but does not specify a methodology to conduct it or consistently provide error ranges for default values. The IPCC should define an explicit approach to approximate the accumulation of errors in multiplicative and additive calculations.

Given the importance of emissions from charcoal production and biofuel combustion, and the fact that few data are now available on these emissions, research should be conducted. Relatively simple equipment and research programs will be sufficient to conduct this research and improve data on these emissions fairly rapidly. Individual national-level studies are not necessary; regional studies targeting a few countries in each region would provide representative regional data.

3.2.3. Mitigation options

- *Savanna Fire Suppression.* Savannas are burned for a variety of purposes, and savanna fires are a necessary part of ecological and agricultural processes. In addition, high rainfall seasonality makes savannas very fire-prone. As a result, complete suppression of fires is not a realistic option. However, some reduction in savanna burning frequency may be a viable mitigation option. This action would have two consequences: it would reduce emissions of GHGs and ozone precursors, and also result in net carbon uptake.

  African countries need to pursue options that will both reduce GHG emissions and be consistent with development goals. Because limiting savanna fires is expected to increase the stock of woody biomass, and because wood is used for a variety of productive purposes (e.g., construction, fuel), this option could have positive development effects. It would, however, have negative long-term effects on grazing potential.
- *Biofuel combustion and charcoal production.* There are many options for reducing GHG emissions through improvements in the efficiency of cookstoves and charcoal production processes. This can have two benefits: limiting direct emissions, and preserving carbon sinks through reduction in biofuel use. Field studies would be necessary to demonstrate these benefits.
- *Slash-and-Burn Agriculture.* New agriculture methods such as agroforestry and alley cropping are being tried, but have not been successful so far.
- *Mitigation costs.* Mitigation options such as forestation (to provide alternative fuel sources), sustainable management of existing indigenous woodlands, and more efficient wood stoves and charcoal production processes could have short-term startup costs, but have low or even negative costs in the long run. The costs of reducing savanna burning may be relatively low as well. Pilot projects to determine costs and results of these mitigation programs would be useful.

3.2.4. Programmatic issues

A few programs have contributed to improving emission estimates (e.g., SAFARI in South Africa for savanna burning data in South and Central Africa, DECAFE for West Africa). A few future campaigns already planned will give more information on emission factors (e.g., the next phases of SAFARI and EXPRESSO in Africa, and SCAR-C in South America). The proposed IGBP megatransects would greatly improve fuel load estimation.

Research priorities should be emissions from biofuels, areal extent of forest and savanna burning, and identification and cost assessment of mitigation options. Research to estimate the effects of limiting savanna fires on biomass accumulation and the ecology and usefulness of savannas is still needed. In the long term, it may also be possible to systematically collect data on fire occurrence at the continental scale with little incremental cost, using AVHRR systems already in place to study land cover.

3.2.5. Conclusions

- *Scope.* Vegetation burning is an important source of GHG emissions in Africa; in many countries, it may be the dominant source. No conclusion is possible on whether savanna burning is increasing or decreasing as an emission source. Slash-and-burn agriculture is likely to peak during the next two decades and then decline. The *IPCC Guidelines* should be expanded to include aerosol and NMHC emissions from vegetation burning.
- *Methods and Data.* The basic IPCC methodology for calculating emissions from savanna burning (i.e., estimating emissions from data on area burned, fuel load, and emission factors) is appropriate. Because better data are available on emission factors than on fuel load and area burned, future research should focus on estimating fuel load and areas. In estimating areas burned, remote sensing (low-resolution calibrated with high-resolution) should be used. Most African fires are anthropogenic, and all fires should be included in emission calculations. For estimating areas and carbon stock for slash-and-burn agriculture, remote sensing in combination with ground teams could be used. Agricultural residue burning is not a significant emission source in Africa, because most residues are used for energy, feed, and other purposes. Good data are available on emission factors for savanna burning, and relatively good data are available on emission factors for other vegetation burning. Vegetation emission factors presented in the *IPCC Guidelines* could be improved based on recent data. Research should be conducted on emissions from charcoal and biofuel production and use.
- *Mitigation.* Because savanna fires serve useful agricultural and ecological purposes, complete suppression is not an option for mitigation; however, it may be possible to reduce fire frequency. Other mitigation options include afforestation, sustainable woodland management, and more efficient cookstoves and charcoal production processes. These options may have low or negative costs in the long run. African countries need to pursue options consistent with their development goals.
- *Programmatic Issues.* Programs that have contributed to improving emission estimates include SAFARI for South and Central Africa and DECAFE for West Africa. Future campaigns, including SAFARI and EXPRESSO in Africa and SCAR-C in South America, will provide more information on emission factors. Research priorities include emissions from biofuels, areal extent of forest and savanna burning, and identification and cost assessment of mitigation options.

3.3. Livestock and Livestock Manure

Co-Chairs: Richard S. Muyungi & Andrew de Jode
Rapporteur: Michael J. Gibbs

3.3.1. Emission inventory scope

Overall, the scope for the livestock and manure sectors is well defined and understood. The activities included and the GHGs addressed are clear. Possible improvements include:

- *Additional GHGs.* Additional GHGs could be investigated related to emissions from manure management. In particular, in several African countries large piggery installations with high concentrations of animals are becoming increasingly common. In addition to releasing CH_4, the liquid manure management facilities at these installations are believed to release CO as well as various nitrogen-containing compounds such as N_2O and NO_x. The significance of these additional GHGs should be examined for potential inclusion in the methods and inventory.
- *Managed Wildlife.* In several African countries, wildlife is considered to be managed, as human intervention helps to maintain the number, balance, and health of wildlife in game reserves, national parks, and on communal and private lands. The CH_4 emissions from managed wildlife should be recognized in the emission inventory, and reported separately from the estimates for domesticated livestock. The IPCC should identify suitable methods for estimating emissions from managed wildlife in Africa, particularly ruminant wildlife.

Within Africa, the livestock sector is extremely important from an economic and social point of view. Most economies are highly dependent on livestock, and in many cases the majority of households keep some form of livestock. Livestock are used not only for milk, meat, and draft power, but also for cultural purposes and as a source of emergency finances. Within the GHG inventories represented among the working group, livestock are by far the most important source of CH_4 emissions. Additionally, livestock are expected to be a significant portion of the total GHG inventory for all gases for many African countries. Therefore, this sector is important in Africa for GHG emission inventories and (possibly) mitigation.

Various factors are leading to changes in emissions from livestock in the region. For example, in some areas animal numbers may be increasing due to the adoption of small scale zero-grazing (i.e., intensive) dairying. Balancing this trend, however, is the expectation that animal numbers may be declining among some pastoralist societies as they become more settled. The working group did not know the net result of the various factors contributing to changes in animal numbers, although FAO data values show a general trend of increasing animal numbers for the region in the most recent decade.

In addition to changing animal numbers, the emission factors for livestock may be changing. On the one hand, the small-scale zero-grazing dairying operations may have higher emissions per animal as each animal has a higher rate of production and hence feed intake. This fact is balanced by the improved diet condition of the animals which tends to reduce CH_4 production per unit of feed intake. The net result of these two factors among this subpopulation of animals has yet to be determined. It is clear, however, that CH_4 emissions from manure management are increasing in the region as liquid manure management is becoming increasingly common among the small-scale zero-grazing dairy operations and large piggery installations discussed above.

3.3.2. Emission inventory methodology and data

Experience in preparing emission inventories suggests the following points:

- *IPCC Method.* Although generally complete and well described, the IPCC method needs to be clarified regarding the use of the more-detailed "Tier 2" equations for estimating emission factors. It needs to be made clear that the equations apply only to cattle, and not to other ruminant animals such as sheep and goats. Additionally, the proper equations for sheep, goats, and other animals should be provided in the methods.
- *Data Verification.* Critical data, such as livestock populations, are generally available from government reports and bulletins. Additional data collection from experts, in particular to characterize the livestock populations, is required. Considerable effort is required to verify the available data, and this verification activity is the most time consuming step of preparing the inventory. To augment the use of experts, limited field data collection may be appropriate to help verify the available data, in particular for those livestock populations that contribute most to the estimated emissions.
- *Feed Characteristics.* Because feed characteristics, particularly digestibility, influence the amount of CH_4 emitted, data describing feed consumed by livestock are needed. Identifying data for this estimate can be time consuming because regional and seasonal differences in feed characteristics need to be reflected. A feeding calendar can be used to identify the types of feed consumed by time of year, and this information can be used to reflect differences in feed characteristics seasonally.

Data collection and verification is the major activity required to conduct the inventory for this sector; the calculations themselves are relatively simple and straightforward. It is valuable to circulate an initial inventory for comment among livestock experts early in the inventory process with the expectation that the experts can provide needed feedback on the proper characterization of the livestock and livestock feeding conditions.

Obtaining data describing manure management practices can be difficult. In Africa, most manure is managed in dry systems (including manure used for fuel and construction), and emissions from these practices are extremely small. However, liquid systems are being used increasingly in small-scale zero-grazing dairies and piggeries and the manure handled at these installations may produce significant amounts of CH_4 in specific countries.

As discussed above, wildlife should be added to the inventory. Data on animal numbers should be available as these numbers are routinely collected for purposes of managing the wildlife. However, default emission factors and suitable equations for estimating "Tier 2" emission factors are required for the types of wildlife commonly managed. The types of wildlife to be included need to be determined based on consultation with experts in this area.

3.3.3. Mitigation options

Livestock and livestock manure emissions are suitable for mitigation in the African context. Mitigation strategies must be developed that correspond to the motivations and resources of those who manage specific sub-sectors of livestock populations. While mitigation opportunities appear to apply broadly to the entire livestock sector, this discussion focuses on the largest source of emissions in Africa, which is cattle. The major sub-sectors of the cattle population are as follows:

- *Dairy production:*
  - large scale commercial enterprises
  - small scale zero-grazing enterprises
- *Beef production:*
  - commercial ranching, often followed by fattening
- *Pastoralist production system:*
  - agro-pastoralist, partially settled and increasingly linked to the monetary economy
  - pure pastoralist, substantially independent from sedentary economic parameters

Within the African context, the dairy and beef sectors, both large and small scale, present opportunities for mitigation. Improvements in production efficiency, linked with infrastructure development and related market access, can be used to reduce enteric fermentation emissions.

The promotion and use of biogas digesters for energy production can be used to reduce the increasing emissions associated with manure management at piggery and small scale dairy installations. In addition, the biogas provides energy for lighting and cooking in the household. As an added benefit, the biogas energy displaces some amount of the household fuelwood requirement, thereby reducing pressures on forest resources which may otherwise contribute to CO_2 emissions as the result of forest degradation.

While there are opportunities to mitigate emissions from these sectors, the pastoralist sub-sector will be the largest source of livestock GHG emissions in many countries. The potential to mitigate emissions from this group is debatable. The following conclusions are possible:

- The extent of pure pastoralist production, generally isolated from sedentary economic parameters, is declining as partial settlement of these populations occurs and as pastoralists become increasingly integrated into the monetary economy.
- The partially settled agro-pastoralists appear to be suitable candidates for mitigation, although care must be taken to design mitigation approaches that recognize the motivations of, and resources available to, this sector.
- Subregional collaboration (e.g., in East Africa) to investigate mitigation options for the pastoralist sub-sector is warranted.

Overall, mitigation in the livestock sector is feasible and promising. However, it must be integrated with overall national development objectives for this sector. As such, the livestock development plans for each country are a useful starting point for identifying the *incremental* efforts appropriate for reducing emissions.

While not strictly barriers in all cases, key features of the livestock situation in Africa require attention as part of mitigation activities. These key features include:

- *Infrastructure.* Livestock emissions mitigation must be addressed from the overall development perspective. As such, one important component of any effort will be to ensure the satisfactory provision of adequate infrastructure to link livestock producers to key inputs as well as to markets. In some cases, existing livestock sector development efforts include infrastructure initiatives. These initiatives must be fulfilled in order for the opportunities for mitigation to be achieved.
- *Coordination with programs with mutual objectives.* There are a variety of development initiatives relating to improving the quality of life of rural people. The livestock sector initiatives will benefit by coordinating with programs that have mutual objectives.
- *Involvement and awareness.* Livestock initiatives require a commitment at all levels, including among policymakers, researchers, and producers. An inclusive process must be undertaken that identifies the mitigation approaches that are most appropriate for local conditions and that will provide a benefit to producers.
- *Overcome initial reluctance.* There is often initial reluctance to adopting new agricultural practices. Involvement and awareness on the part of producers during the formulation of mitigation options is necessary to lessen this reluctance. However, there is also a role for government initiatives to help overcome this reluctance. Avenues for accomplishing this objective may include, among others: conducting demonstration projects within key production districts; providing information and training programs; and providing temporary incentives for early adopters of new production techniques to offset potential risks.

3.3.4. Programmatic issues

A variety of national and international programs have been very important in the development of national inventories. Among the international programs, the *IPCC Guidelines* program has made a critical contribution to the overall consistency and quality of inventory efforts globally. Among funding organizations, the U.S. Country Studies Program, the UNEP/GEF Country Case Studies Project, and other key programs have enabled the countries to prepare materials as called for under the Framework Convention on Climate Change. In particular, the training components of these programs have been enormously valuable.

Several national programs are important for the completion of the inventory for the livestock sector. Ongoing efforts in various countries to develop a national livestock census are, of course, of particular significance. Additionally, Agriculture Ministries, or similar organizations, often have periodic bulletins relating to livestock conditions and production practices. Finally, production experts at research institutions or within the extension service are important for verifying data and providing advice on the characterization of production practices.

There are several gaps in the available programs:

- *Networking.* Improved networking among those preparing inventories would be extremely helpful so that data and insights could be shared. The workshop format is a particularly good method for communicating on these issues. Subregional workshops should be conducted; within the livestock sector there are very strong similarities in animal types and production systems within subregions, while there are larger differences among the subregions.
- *Regional Collaboration.* Regional collaboration can help develop key data required for emission inventories and mitigation. For example, region-specific or sub-region-specific emission factors need to be verified. A regionally-coordinated emission measurement program should be a high priority.
- *Involvement.* Increased participation by livestock experts in Africa is needed. Improved links should be developed between these experts and those conducting emission inventories. Increasing awareness and understanding of the links between the livestock sector and environmental issues is needed.

3.3.5. Conclusions

- *Scope.* Within Africa the livestock sector is extremely important from an economic and social point of view. Livestock appear to be responsible for a significant portion of the total GHG emissions for many African countries, making the sector important in Africa for both emission inventories and mitigation.

  The scope for the livestock and manure sectors is well defined and understood. Recommendations include: (1) additional GHGs from manure management could be investigated; and (2) CH_4 emissions from managed wildlife should be recognized in the emission inventory.
- *Methods and Data.* The IPCC methods could be clarified, although overall the methods are satisfactory. Data collection and verification are the most challenging aspects of conducting the inventory for this sector, and regional networking can provide useful discussions of data issues. Africa-specific emission factors must be developed and verified to ensure that the methods are adequate for the African context.
- *Mitigation Options.* Livestock and livestock manure emissions are suitable for mitigation in Africa. Mitigation strategies must be developed that correspond to the motivations and resources of managers of specific sub-sectors of the overall livestock populations. In particular, livestock emissions mitigation must be integrated with overall national development objectives for this sector. As such, the livestock development plans for each country are a useful starting point for identifying the *incremental* efforts appropriate for reducing emissions.
- *Programmatic Issues.* A variety of international and national programs are important for supporting the livestock emission inventory and mitigation analyses. There are three primary improvements that should be made in the available programs: (1) Improved networking among those preparing inventories would be extremely helpful so that data and insights could be shared. (2) Regional collaboration could help develop key data

required for emission inventories and mitigation. A regionally-coordinated emission measurement program should be a high priority. (3) Increased participation by livestock experts and producers in Africa is needed. Improved links should be developed among these individuals and those conducting emission inventories and mitigation analyses.

References

UNEP/OECD/IEA/IPCC: 1995, *IPCC Guidelines for National Greenhouse Gas Inventories*, IPCC, Bracknell, 3 Volumes.

Scholes, R.J.: 1995, Estimating greenhouse gas emissions from vegetation fires in Africa, *Environmental Monitoring and Assessment*, in press.

PART II

INTERNATIONAL CONTEXT

SCIENTIFIC ASPECTS OF THE FRAMEWORK CONVENTION ON CLIMATE CHANGE AND NATIONAL GREENHOUSE GAS INVENTORIES

B. CALLANDER

Technical Support Unit for IPCC WGI (Science Assessment), Hadley Centre for Climate Prediction and Research, Meteorological Office, Bracknell, UK.

Abstract. Though the principles of the Earth's greenhouse effect have been known for well over a century, it is only recently that advances in climate research have indicated that significant and possibly costly climate change, due to growing emissions of greenhouse gases and their precursors by human activity, is a real possibility. Current estimates of the global human-related emissions of carbon dioxide, methane and nitrous oxide are presented, though many sources remain poorly known or understood. The compilation of national greenhouse inventories as required by the United Nations Framework Convention on Climate Change is likely in the longer term to help improve such global estimates, as long as comparable methodologies are used. The development of the *IPCC Guidelines for National Greenhouse Gas Inventories* is described, emphasizing the strategies employed to gain wide international participation.

1. Introduction

Every party to the United Nations Framework Convention on Climate Change (FCCC) is obliged to submit and regularly update a national inventory of greenhouse gas emissions by sources and removals by sinks. Technical experts tasked with compiling such inventories are likely to be more highly motivated if they have confidence in the methods they are asked to use, and if the work has more than national relevance and fits into a much larger global picture. The purpose of this paper is to clarify some of the scientific issues underpinning the FCCC, to underline the importance of national greenhouse gas inventories not only for the individual country itself, but for the overall purposes of the Convention, and to describe important elements in the development of the *IPCC Guidelines for National Greenhouse Gas Inventories.*

2. Scientific background: the enhanced greenhouse effect

Some gases present in our atmosphere absorb long-wave radiation emitted by the surface of the earth and by clouds. The presence of these so-called greenhouse gases in the atmosphere keeps the earth substantially warmer—approximately thirty degrees Celsius warmer—than it would otherwise be. This is the natural greenhouse effect (Figure 1). Important natural greenhouse gases are water vapor (H_2O), carbon dioxide (CO_2), methane (CH_4), and ozone (O_3). The major components of the atmosphere, nitrogen (N_2) and oxygen (O_2), do not interact with long-wave radiation and play no part in the greenhouse effect.

Environmental Monitoring and Assessment **38**: 129-140, 1995.

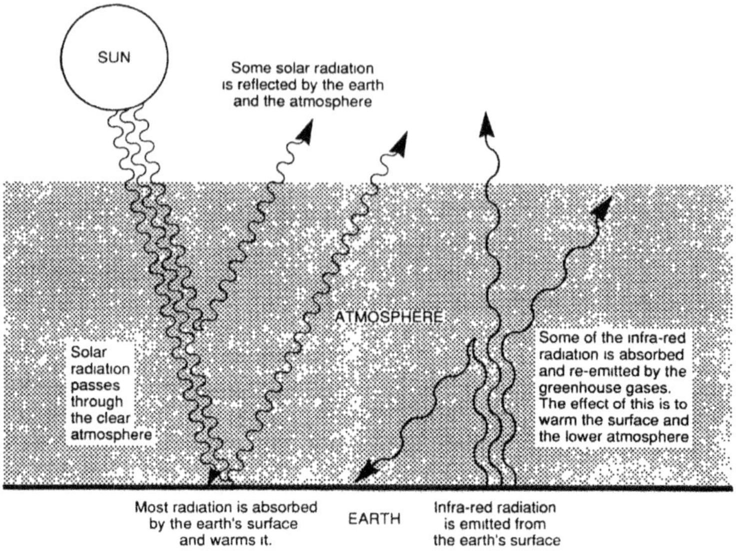

Figure 1. The natural greenhouse effect (IPCC, 1990).

The atmospheric concentrations of CO_2, CH_4, and N_2O have apparently been maintained for centuries at constant levels by a balance of natural fluxes into and out of the atmosphere.

Human activities are adding greenhouse gases to the atmosphere in increasing amounts. The distribution and abundance of water vapor is controlled by natural factors such as the temperature structure of the atmosphere and are not directly affected by human activities. In contrast, the atmospheric concentrations of CO_2, CH_4, and other greenhouse gases are now rising as a direct result of human activities (Figure 2). The concentration of new gases such as the chlorofluorocarbons (CFCs), not present in the pre-industrial atmosphere, have also increased. The effect of this is to alter the radiative energy balance of the earth-atmosphere system. The rise of greenhouse gas concentrations since pre-industrial times is calculated to have added around 2.5 Wm^{-2} to the net energy entering the earth-atmosphere system. This may seem a small amount, but it is probably greater (in terms of century-long averages) than any perturbation to the system from natural sources within the last several thousand years of the earth's history (Figure 3).

The belief that additional greenhouse gases in the atmosphere affect the energy balance of the earth-atmosphere system and that the climate must respond in *some* way to this perturbation, rests on well-established physics and is largely beyond dispute. In fact, the warming effect of greenhouse gases in the atmosphere was first recognized in 1827 by Jean-Baptiste Fourier, and almost a century ago a Swedish scientist, Arrhenius, made the first calculation of the effect of doubling atmospheric concentrations of CO_2. His estimate,

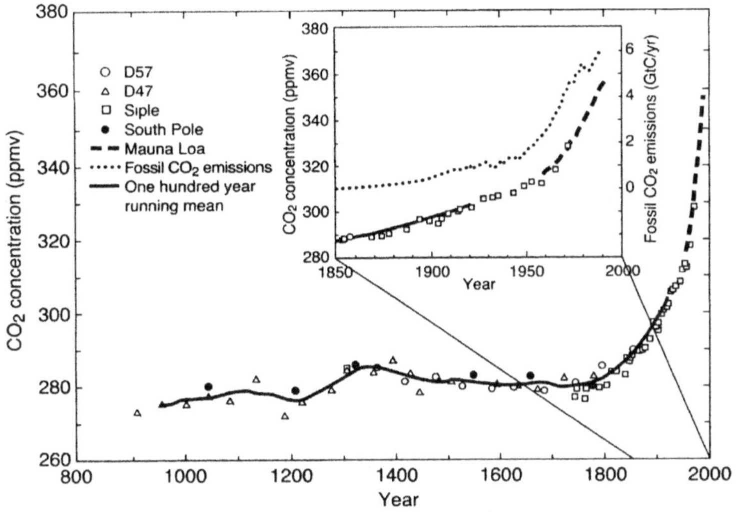

Figure 2. Carbon dioxide concentrations over the last 1000 years (IPCC, 1994).

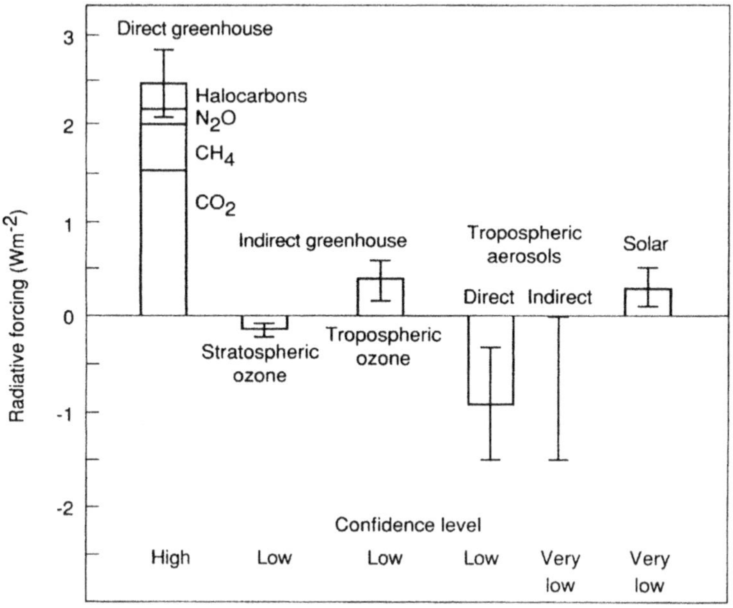

Figure 3. Comparison of the relative "climate-forcing" effects of gases related to human activity and solar variation, a natural factor (IPCC, 1994).

a global average temperature increase of 5 to 6 degrees Celsius, is not very different from current estimates using much more sophisticated techniques.

Arrhenius himself apparently recognized that calculation of a global average temperature rise was not in itself particularly useful, because the impacts of global warming would be determined by climate change at the *regional* level. But the mathematical tools he needed to address this question were not available in his day, and it has been only within the last twenty years or so that numerical models of climate, based on our physical understanding of climate processes and requiring very powerful computers, have been able to provide scenarios of possible future climate change at the regional level.

3. History of the Intergovernmental Panel on Climate Change and the United Nations Framework Convention on Climate Change

During the 1980s the idea of "global warming" received much exposure by the media, who often presented their stories in association with lurid images of disaster or of a climate out of control. This left the public and governments disturbed but not necessarily better informed. In response to this general concern and the lack of a clear, objective, authoritative statement on the problem on which governments might base policy, the World Meteorological Organisation (WMO) and the United Nations Environmental Programme (UNEP) jointly established in 1988 the Intergovernmental Panel on Climate Change (IPCC), charged with the "assessment of available scientific information on climate change, ... of the environmental and socio-economic impacts ... and formulation of response strategies."

The IPCC has been remarkably successful in engaging the world's leading experts in the many fields related to the science of climate and climate change, to its impacts, and to possible options for adaptation and mitigation. For example, the first IPCC Scientific Assessment, submitted to the Second World Climate Conference in 1990, combined contributions from 170 scientists from 25 countries and was reviewed by a further 200 scientists. The number of scientists involved in later IPCC assessment has been even greater. The first IPCC reports of 1990 acknowledged the large degree of uncertainty still associated with the climate change issue. Nevertheless, the overall message—that many human and natural systems are vulnerable to the magnitudes and/or rates of climate change likely over the next century—added impetus to the negotiation of the UN Framework Convention on Climate Change, which was eventually signed in June 1992 in Rio de Janeiro, Brazil, by more than 150 countries.

4. Greenhouse gas emissions: the global picture

4.1. Carbon Dioxide

There are three major sources of CO_2 emissions linked to human activity: fossil fuel burning, cement production, and forest harvest and other changes of land use. IPCC (1994) assessed the magnitude of these sources (Table I). Though relatively small when compared with the natural fluxes of carbon between the terrestrial and marine biospheres and the atmosphere, these sources are sufficiently large to upset the balance between the natural fluxes, resulting in a relatively rapid rise in atmospheric concentrations of CO_2 since the beginning of the industrial revolution (Figure 2). Atmospheric concentrations (averaged over several years) are currently rising at about 1.5 ppmv (0.4%) per year.

4.2. Methane and Nitrous Oxide

Human-related emissions of CH_4 and N_2O represent a large proportion of total emissions of these gases, though there is substantial uncertainty over the magnitude of the natural sources (Tables II and III).

4.3. The Importance of National Inventories

Tables I, II, and III contain estimates of *global* emissions, and it is very clear that there is considerable uncertainty about some of the numbers. Compilation of greenhouse gas inventories will, in due course, allow the uncertainties to be reduced, because of the greater amount of information that will arise from detailed consideration of the factors that govern emissions in different regions and under different local conditions. Without such efforts, estimation of the strength of many of the sources and sinks contributing to global anthropogenic emissions would remain incomplete.

TABLE I

Average annual CO_2 budget for perturbations to the carbon cycle by human activities for the period 1980 to 1989. Error limits correspond to an estimated 90% confidence interval.

	Billion tonnes of carbon/year
CO_2 sources	
(1) Emissions from fossil fuel combustion and cement production	5.5 ± 0.5
(2) Net emissions from changes in tropical land use	1.6 ± 1.0
(3) Total anthropogenic emissions (1) + (2)	7.1 ± 1.1
Partitioning among reservoirs	
(4) Storage in the atmosphere	3.2 ± 0.2
(5) Oceanic uptake	2.0 ± 0.8
(6) Uptake by Northern Hemisphere forest re-growth	0.5 ± 0.5
(7) Additional terrestrial sinks (CO_2 fertilization, nitrogen fertilization, climatic effects) [(1) + (2)] − [(4) + (5) + (6)]	1.4 ± 1.5

From IPCC, 1994.

TABLE II

Estimated natural and anthropogenic sources of CH_4, in millions of tonnes of CH_4 per year

Identified sources	Individual estimate	Total
Natural		
Wetlands	115 (55–150)	
Termites	20 (10–50)	
Oceans	10 (5–50)	
Other	15 (10–40)	
Total identified natural sources		160 (110–210)
Anthropogenic		
Individual fossil-fuel-related sources		
Natural gas	40 (25–50)	
Coal mines	30 (15–45)	
Petroleum industry	15 (5–30)	
Coal combustion	? (1–30)	
Total fossil-fuel related		100 (70–120)
Biospheric carbon		
Enteric fermentation	85 (65–100)	
Rice paddies	60 (20–100)	
Biomass burning	40 (20–80)	
Land fills	40 (20–70)	
Animal waste	25 (20–30)	
Domestic sewage	25 (15–80)	
Total biospheric		275 (200–350)
Total identified anthropogenic sources		375 (300–450)
Total identified sources		535 (410–660)

From IPCC, 1994.

But the value of a national greenhouse gas inventory is much more than the improvement of global estimates. First, it allows each country to compare its emissions with those of others. Second, it provides for each country a baseline against which to judge the success of greenhouse gas mitigation policies. Third, by breaking down emissions according to category, it also provides guidance on where greenhouse gas emissions might be controlled most effectively and efficiently.

TABLE III

Estimated sources of N_2O typical of the last decade, in millions of tonnes of nitrogen per year

Identified sources	Range	Likely
Natural		
Oceans	1–5	3
Tropical soils		
Wet forests	2.2–3.7	3
Dry savannas	0.5–2.0	1
Temperate soils		
Forests	0.1–2.0	1
Grasslands	0.5–2.0	1
Total identified natural sources	6–12	9
Anthropogenic		
Cultivated soils	1.8–5.3	3.5
Biomass burning	0.2–1.0	0.5
Industrial sources	0.7–1.8	1.3
Cattle and feed lots	0.2–0.5	0.4
Total identified anthropogenic	3.7–7.7	5.7
Total identified sources	10–17	14.7

From IPCC, 1994.

5. The relative importance of emissions of different greenhouse gases

Three factors are important to the ways gases vary in their contribution to the enhanced greenhouse effect.

The first factor is the ability of the gas to absorb radiation. A molecule of CO_2 absorbs less radiation than a molecule of CH_4 or N_2O, which in turn absorb less radiation than a molecule of CFC–12, one of the halocarbons (Table IV).

The second factor is the lifetime of the gas in the atmosphere, because climate change is a result of a *sustained* alteration of the Earth's energy balance over a period of time. All anthropogenic greenhouse gases eventually react with other atmospheric constituents or are decomposed by ultraviolet radiation. The only exception is CO_2, which is removed from the atmosphere by processes at the Earth's surface, photosynthesis being the most important. The rates of atmospheric removal vary among the different gases, leading to different atmospheric lifetimes for different gases (Table V).

TABLE IV

Increase in radiative forcing due to the addition of equal masses of some important greenhouse gases to the atmosphere. All numbers are expressed relative to CO_2, and assume present day concentrations in the atmosphere.

Trace gas	Increase in forcing per unit mass relative to CO_2
Carbon dioxide, CO_2	1
Methane, CH_4	58
Nitrous oxide, N_2O	206
CFC-12 (an ozone-depleting gas controlled under the Montreal Protocol and its Amendments)	5750
HFC-125 (a possible substitute for the CFCs)	4920

Derived from Table 2.3 of IPCC, 1990.

TABLE V

Some Global Warming Potentials

Species	Lifetime (Years)	Global Warming Potential over specified time horizon		
		20 years	100 years	500 years
Carbon dioxide, CO_2	~100	1	1	1
Methane, CH_4	14.5	62	24.5	7.5
Nitrous oxide, N_2O	120	290	320	180
For comparison, a gas with a very large GWP, but which currently contributes relatively little to climate change because emissions are relatively small				
Sulphur hexafluoride, SF_6	3200	16 500	24 900	36 500

From IPCC, 1994.

Factors 1 and 2 are combined in an index termed the Global Warming Potential, which allows the warming impact of equal emissions of different gases, over specific time horizons, to be compared (Table V).

The third factor is the total mass of gas added to the atmosphere. A relatively weakly absorbing gas (on a per molecule basis) can be a globally important greenhouse gas if sufficient quantities are emitted. The relative importance of different gases to the enhanced greenhouse effect is obtained by multiplying the GWP value by the total mass of gas emitted (Figure 4). Thus, because of the very large global amounts of CO_2 emitted through human activities, this gas is calculated to make the dominant contribution to the enhanced

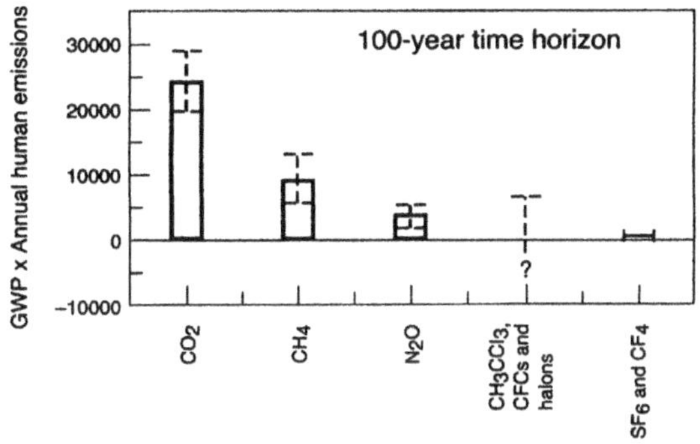

Figure 4. Product of GWP with estimated annual global anthropogenic emissions typical of the 1980s for various gases, for a 100-year time horizon. From IPCC, 1994.

greenhouse effect even though, on a per molecule basis, it seems to be a relatively weak greenhouse gas.

The products shown in Figure 4 are for global emissions, but the relative importance of different greenhouse gases for a particular country is obtained in a similar way, by multiplying the national emissions of each gas by the respective GWP.

6. National inventories of greenhouse gas emissions and the Climate Convention

6.1. The Terms of the UN Framework Convention on Climate Change

The ultimate objective of the UNFCCC is expressed in Article 2:

> The ultimate objective of this Convention and any related legal instruments that the Conference of the Parties may adopt is to achieve, in accordance with the relevant provisions of the Convention, stabilization of greenhouse gas concentrations in the atmosphere at a level that would prevent dangerous anthropogenic interference with the climate system. Such a level should be achieved within a time frame sufficient to allow ecosystems to adapt naturally to climate change, to ensure that food production is not threatened and to enable economic development to proceed in a sustainable manner.

At the moment it is not possible to define what level of stabilization the Convention should aim for, because it is not yet possible to relate the sorts of "dangerous interference" listed in Article 2 to particular concentrations of greenhouse gases.

What we *can* calculate with some confidence is that present levels of *emissions* are likely to produce increasing atmospheric *concentrations* of greenhouse gases for years, decades, or even centuries into the future. Whatever level of stabilization is eventually agreed upon, reduction, control or, at the very least, planning of global greenhouse gas emissions is likely to form a key component of future agreements negotiated under the Convention. Article 4 (Commitments) recognizes that the compilation of national inventories of current emissions is a necessary and fundamental component of that process:

> 1. All Parties, taking into account their common but differentiated responsibilities and their specific national and regional development priorities, objectives and circumstances, shall:
> (a) Develop, periodically update, publish and make available to the Conference of the Parties ... national inventories of anthropogenic emissions by sources and removals by sinks of all greenhouse gases not controlled by the Montreal Protocol, using comparable methodologies to be agreed upon by the Conference of the Parties;
> (b) Formulate ... measures to mitigate climate change by addressing anthropogenic emissions ... and removals ... of greenhouse gases not controlled by the Montreal Protocol ...

The phrase in the preamble to Article 4, "*common but differentiated responsibilities*", reflects an important principle of the Convention, allowing different countries ("Parties") to follow different timetables and to bear different responsibilities depending on their circumstances. Thus the industrialized countries (known in the Convention as "Annex I countries") were required to submit their national greenhouse inventories by September 1994, six months after entry into force of the Convention, whereas many other countries have until 1997, or even longer, to do so.

The IPCC *Guidelines for National Greenhouse Gas Inventories* (UNEP *et al.*, 1995) is so far the only inventory methodology formally agreed by the Conference of the Parties for use by all countries, though Parties are permitted to use "*an established and comparable methodology ... provided that they include sufficient documentation to back up the data presented.*"

6.2. The IPCC/OECD/IEA Program to Develop *IPCC Guidelines for National Greenhouse Gas Inventories*

The current IPCC *Guidelines* (UNEP *et al.*, 1995) were developed over three years (1991 to 1994) in a joint program between IPCC, the Organisation for Economic Co-operation and Development (OECD), the International Energy Agency (IEA), and UNEP. The program aimed to produce a methodology having three key features.

The first feature is flexibility and comprehensiveness in scope. The Phase I *Guidelines* provide a default methodology covering the known sources and sinks of the most important anthropogenic greenhouse gases. However the *Guidelines* can also accommodate other methodologies that use different levels of detail and sophistication. The *Guidelines* provide instructions for converting results of other inventory approaches into a comparable reporting format and for documenting the methods and data used to allow verification of consistency. The *Guidelines* recognize that varying levels of uncertainty are associated with different parts of the methodology and provide recommendations for managing, reporting, and minimizing uncertainty.

The second feature is the establishment of a strong technical pedigree for each component of the methodology. This was achieved by convening, for each gas and for each sector, expert groups drawn from as wide a geographical base as possible, whose purpose was to make recommendations for the methodology based on a critical assessment of the available scientific literature. Before receiving formal IPCC approval in September 1994, the draft *Guidelines* were submitted to a formal IPCC intergovernmental review process involving individual experts and governments of the member countries of IPCC.

The third feature is widespread and effective participation by the "user community" of many countries in the development of the *Guidelines*. A network of national technical points of contact covering more than sixty countries was established and used as a conduit for the dissemination and testing of draft material, including computer software. Various international workshops were held in order to expose the draft methodology to international review and comment, and to examine issues involved in developing a common reporting format for inventories compiled under different methodologies. In addition, country studies funded by UNEP under the Global Environment Facility were carried out in ten developing countries. The aim of each of these studies was to prepare a national emissions inventory according to the draft IPCC methodology and, through this exercise, to gain much additional information on the operation of the methodology in different national contexts.

It is important to recognize, however, that development of the *Guidelines* is an ongoing task and that publication of the Phase I *Guidelines* does not represent an end point. The UNEP/OECD/IEA/IPCC program is currently engaged in a second phase of activity, due to be completed around mid-1996, the outcome of which will be recommendations for further improvements and additions to the existing Phase I *Guidelines*. The present Johannesburg workshop can make an important contribution to the Phase II program by providing feedback on the strengths and weaknesses of the IPCC methodology when applied to real situations. This is particularly valuable for emissions from land use change, a sector that presents particular problems when trying to establish methods of general applicability.

7. Conclusion

The climate change issue is likely to be with us for a long time, first because the global climate is extremely complex and improved understanding and predictive capability are likely to emerge only slowly, and second because the trends of increasing atmospheric concentrations of greenhouse gases are unlikely to be reversible in time scales that are short compared to a human life span. National inventories can play an important part in understanding how greenhouse gas emissions from human activity have evolved in the past and in projecting how they might evolve in the future.

References

IPCC (Intergovernmental Panel on Climate Change): 1990, *Climate Change: the IPCC Scientific Assessment*, Houghton, J.T., Jenkins, G.J., and Ephraums, J.J. (eds.), Cambridge University Press.

IPCC: 1994, *Climate Change 1994: Radiative Forcing of Climate Change and Evaluation of the IPCC IS92 Emission Scenarios*, Houghton, J.T., Meira Filho, L.G., Bruce, J., Lee, Hoesung, Callander, B.A., Haites, E., Harris, N., and Maskell, K., (eds.), Cambridge University Press.

UNEP, OECD, IEA, IPCC (United Nations Environment Programme, Organisation for Economic Co-operation and Development, International Energy Agency, Intergovernmental Panel on Climate Change): 1995, *IPCC Guidelines for National Greenhouse Gas Inventories*, IPCC, Bracknell, 3 Volumes.

SELECTED INTERNATIONAL EFFORTS TO ADDRESS CLIMATE CHANGE

M. SEKI and R. CHRIST

Atmosphere Unit, United Nations Environment Programme (UNEP), P.O. Box 47074, Nairobi, Kenya

Abstract. Over the past two decades, concern about human-induced climate change has become an increasingly important item on the environmental and political agenda. The signing of the United Nations Framework Convention on Climate Change and the adoption of Agenda 21 at the United Nations Conference on Environment and Development in Rio de Janeiro in 1992 provided international organizations and the nations of the world with a new focus for climate-related activities. Although there remains considerable scientific uncertainty about the extent, magnitude, and rate of climate change and the impacts of such change, actions to address climate change have been initiated both internationally and nationally. Major international activities include the World Climate Programme, the Intergovernmental Panel on Climate Change, the United Nations Framework Convention on Climate Change, and the United Nations Environment Programme.

1. Introduction

The atmospheric environment is threatened by anthropogenic emissions of pollutants and greenhouse gases (GHGs) to the extent that irreversible changes to the climate, the ozone layer, and the quality of the air we breathe could occur. Over the past two decades, concern about human-induced climate change has increased both nationally and internationally. The signing of the United Nations Framework Convention on Climate Change (UNFCCC) and the adoption of Agenda 21 at the United Nations Conference on Environment and Development (UNCED) in Rio de Janeiro in 1992 marked a turning point in international efforts to protect the atmosphere and provided a new focus for the climate-related activities of international organizations and the nations of the world.

This paper provides an overview of some of the most significant international actions to address climate change, followed by a discussion of the United Nations Environment Programme's (UNEP's) activities in the field of climate change, focusing on the country studies projects and networking efforts.

2. A brief background on international actions on climate change

2.1. World Climate Programme (WCP)

The First World Climate Conference was held in February 1979. It adopted "The Declaration of the World Climate Conference" (WMO, 1979) which urged the nations of the world to take full advantage of the present knowledge about climate, to take steps to improve that knowledge, and to foresee and to prevent man-made changes in climate that

Environmental Monitoring and Assessment **38**: 141-153, 1995.

are judged to be adverse. In accordance with the recommendation of the conference, the World Meteorological Organization (WMO), in collaboration with UNEP and the International Council of Scientific Unions (ICSU) initiated the World Climate Programme (WCP) to address the potentially serious global problem of human-induced impacts on the climate system and to promote the proper use of climate information for national socioeconomic development. The principal objectives of the WCP are to assist countries in the application of climate knowledge; and to provide governments, industry, and the general public with information on potential climate variation and change. The WCP was formally launched by the Eighth World Meteorological Congress in April and May 1979.

The main objectives of the WCP have remained generally unchanged during the years of its implementation. The WCP identified global goals shared by nations and established a framework of international coordination and cooperation in the effort to implement climate-related work. The WCP is organized into four subprograms, which are described in the following sections.

2.1.1. The World Climate Data and Monitoring Programme (WCDMP)

The purpose of this component of the WCP is "to improve the availability of reliable data for the purposes of the WCP." Improving the availability of data involves maintenance and, as necessary, enhancement of many kinds of observational networks as well as facilities for data processing and exchange. Furthermore, climate data must be analyzed and presented in a form that enables documentation of the worldwide climate, assessments of climate variability and trends, and applications in resource management and socioeconomic activities.

The WCDMP is organized into five projects:

- A project that uses computer systems for climate data management, including activities such as CLICOM (Climate Computer systems) and INFOCLIMA (the Climate Data Referral Systems) development, use, application, and training
- DARE (Data Rescue), which aims at assisting countries in management, preservation, and use of climatic data for their own territories
- The Climate Change Detection Project, which promotes the use of standard procedures in collecting climate data to ensure that analyses of climate trends and variability are reliable
- The Climate System Monitoring Project, which involves compiling information on large-scale climatic fluctuations, which is then disseminated through monthly bulletins and biennial reviews
- The Biennial Global Climate Reviews, which focus on describing climate system anomalies in a historical perspective, in concert with information on general climatological and climate change issues

When viewed in their entirety, these five projects provide a coordinated, complete, climate data management program to strengthen capabilities, to improve interpretation of climate data bases, and to bridge the gap between developed and developing countries.

WMO is responsible for implementing the WCDMP. UNEP, the United Nations Educational, Scientific, and Cultural Organization (UNESCO) and its Intergovernmental Oceanographic Commission (IOC), ICSU, and the Food and Agriculture Organization of the United Nations (FAO) participate and support the program.

2.1.2. The World Climate Applications and Services Programme (WCASP)

This component of the WCP was established in recognition of the importance of using climate information to reduce society's vulnerability to climate extremes. WCASP also recognizes the need for improvements in the data and in the expertise needed in providing climate information services, particularly in developing countries. WCASP aims to promote applications of existing climate information to the priority areas of food, water, energy, planning and management of land use, urban areas, forests, oceans and coastal zones, buildings, and other sectors such as human health, transport, and tourism. Emphasis is placed on methods of adapting to and mitigating adverse impacts of climate and climate variations. This effort is aimed at improving awareness of the potential benefits of climate application and services as well as improving access to techniques for applying climate information and knowledge.

WCASP is implemented through a number of sectoral projects relating to food production, water resource management, energy production and consumption, and human settlements. These projects address issues of development of applications and services techniques in response to user requirements; assistance to members in developing their climate application services; and development of methodologies to assess climate and climate change effects on various socioeconomic activities.

WMO assumes the lead responsibility for the WCASP. A number of water-related projects are implemented under the heading WCP-Water; these are coordinated by WMO and UNESCO. Several other international agencies and organizations cooperate and interact with the WCASP on these projects.

2.1.3. The World Climate Research Programme (WCRP)

The objective of the WCRP is to provide the organizational framework for international cooperative research projects aiming to understand the physical climate system and the causes of climatic variability and climate change. This requires a quantitative understanding of the four main components of the physical climate system: the global atmosphere, the world ocean, the cryosphere (comprises the continental ice-sheets/caps, mountain glaciers, and sea-ice), and the land surface. National and international contributions are made towards various activities that are organized into manageable multidisciplinary projects and experiments.

The WCRP comprises seven major projects: Climate Model Development, Global Energy and Water Cycle Experiment (GEWEX), Tropical Ocean and Global Atmosphere (TOGA), World Ocean Circulation Experiment (WOCE), Arctic Climate System Study (ACSYS), Stratospheric Processes and their Role for Climate (SPARC), and Climate Variability and Predictability (CLIVAR). The WCRP also cosponsors the System for Analysis, Research, and Training (START) with the International Geosphere-Biosphere

Programme (IGBP) and the Human Dimensions of Global Environmental Change Programme (HDP) of the International Social Science Council (ISSC).

WMO, ICSU, and IOC of UNESCO are responsible for the WCRP.

2.1.4. The World Climate Impact Assessment and Response Strategies Programme (WCIRP)
Initially, this program was called the World Climate Impact Studies Programme (WCIP), the overall objective of which was to assess the sensitivity of socioeconomic sectors to climate variability and change. Its name was changed in 1991, when the task of developing options for socioeconomic strategies to respond to climate variability and change was added to the program.

WCIRP's main projects include: (1) testing methodologies for assessing the impacts of climate change and sea-level rise; (2) promoting and improving the coordination of national climate impact and response strategy programs; (3) improving techniques for making inventories of sources and sinks of GHGs; (4) developing national strategies for responding to climate fluctuations and change; (5) improving dissemination of accurate, complete, and timely information to governments and the public; and (6) assessing air quality and air pollution mitigation strategies (UNEP/Atmosphere, 1995).

UNEP is responsible for implementing the WCIRP. Activities are undertaken in cooperation with relevant agencies, including WMO, ICSU, FAO, UNESCO and its IOC, the Intergovernmental Panel on Climate Change (IPCC), and the UNFCCC Secretariat, and jointly with other components of the WCP.

Research in the 1980s showed that some aspects of climate could be predicted and that human activity could have a serious impact on the global climate system. Much of this research was carried out under the WCP framework, whose investigations of the implications of changing climate indicated a cause for concern.

Over the past 15 years much progress has been made in understanding the global climate system. The WCP and its associated activities have generated substantial knowledge, which provided a foundation for much of the IPCC's work, which in turn provides the scientific basis for the UNFCCC and its implementation. Under the framework of the WCP, significant progress and improvements have been made in obtaining, preserving, standardizing, and exchanging data from the atmosphere, ocean, and land surfaces; in producing and distributing analyses of climate data and applying the resulting information in various socioeconomic sectors; in developing methodologies for assessing the socioeconomic impacts of climate variability and change; and in implementing global and regional experiments that help further the understanding of the climate system and the ability to predict its behavior. The WCP has also stimulated national climate-related activities and enhanced national efforts in climate research. It has also aided in establishment and implementation of National Climate Programmes, which are structured national programs of climate-related activities.

In April 1993, an Intergovernmental Meeting (IGM) on the World Climate Programme was held (WMO, 1993) to review its activities. At the IGM, the WCP was changed to more effectively contribute to the implementation of Agenda 21 and the UNFCCC. The

WCP should also work to further the goals of the United Nations Commission on Sustainable Development and the IPCC as well contribute to the socioeconomic development of countries. The IGM concluded that while substantial achievements had been made by the WCP, its new and evolving requirements could only be met if climate-related activities were better integrated and prioritized. The IGM called for a review of existing international activities and for preparation of a proposal to point the way forward for the WCP. Thus, the international organizations involved in the implementation of the WCP have been working to develop a "Climate Agenda" (FAO, ICSU, UNEP, UNESCO and its IOC, WMO, 1995)—a proposal for an integrated framework of international climate-related programs, which focuses on the needs of governments and takes into account the requirements that exist as a result of various international agreements for protection of the atmosphere.

The Climate Agenda follows the four thrusts suggested for the WCP by the IGM:

1. New frontiers in climate science and prediction
2. Climate services for sustainable development
3. Studies of climate impact assessments and response strategies to reduce vulnerability
4. Dedicated observations of the climate system

The climate agenda has been prepared and has been approved by the WMO Congress and the UNEP Governing Council.

2.2. Intergovernmental Panel on Climate Change

In 1988, WMO and UNEP jointly established the IPCC to formally assess the level understanding about climate change, the environmental and socioeconomic implications of such change, and the possible response options available to governments.

The IPCC First Assessment Reports were completed in August 1990 (IPCC, 1990). In 1992, the IPCC and its three working groups updated their 1990 reports (IPCC, 1992), addressing the key conclusions in light of new data and analyses in an effort to provide the UNFCCC as well as UNCED with the latest information on climate change.

In November 1992, recognizing the need for additional information to support and facilitate ongoing negotiations under the Climate Convention, the IPCC adopted a new structure for the working groups. The tasks of the working groups were changed to the following:

- Working Group I—to assess the available information on the science of climate change, particularly changes arising from human activities.
- Working Group II—to assess the available technical, environmental, social, and economic information on the impacts of climate change and on response options that would enable countries to adapt to and mitigate climate change.
- Working Group III—to deal with cross-cutting economic and other issues related to climate change, including the development and assessment of emission scenarios.

Under this new structure the IPCC is undertaking a Second Assessment. The first task in the Second Assessment was to produce a Special Report of the IPCC (IPCC, 1994) for consideration by the First Conference of the Parties to the UNFCCC (COP–1). The Special Report consisted of the following:

1. Radiative Forcing of Climate Change (1994) with a Summary for Policymakers
2. IPCC Technical Guidelines for Assessing Climate Change Impacts and Adaptation with a Summary for Policymakers
3. An Evaluation of the IPCC IS92 Emission Scenarios with a Summary for Policymakers and a Technical Summary
4. IPCC Guidelines for National Greenhouse Gas Inventories

The IPCC Second Assessment Reports are expected to be completed by the end of 1995.

The IPCC assessments are the result of a coordinated effort by hundreds of experts worldwide, under the Chairmanship of Professor Bert Bolin. All member states of the United Nations are members of the IPCC and have the right to participate in all IPCC decisionmaking. The decisions are normally made during the plenary sessions of the IPCC, to which all members are invited. Intersessional decisions are made by a twenty-eight-member bureau established in Harare in November 1992. The bureau's members are selected in a manner designed to ensure an equitable geographic balance. They represent thirteen developed countries, fourteen developing countries, and one country with an economy in transition.

The IPCC's reports present an unbiased scientific assessment of the climate system, its propensity for change, the impacts of such changes, and the range of responses that can be used to mitigate or adapt to such changes. The IPCC assessments constitute the most authoritative statement on climate change available and have become a standard reference for all those concerned with the issue of climate change—policymakers, scientists, and other experts.

2.3. United Nations Framework Convention on Climate Change

The UNFCCC was adopted by the Intergovernmental Negotiating Committee on May 9, 1992 and signed by governments at the United Nations Conference on Environment and Development (UNCED) in Rio de Janeiro in June 1992 (United Nations, 1992). Over 150 states and the European Community signed the Convention, reflecting widespread recognition that climate change is a problem that can only be effectively addressed through international cooperation. Upon ratification by fifty states, the Convention entered into force on March 21, 1994. As of July 31, 1995, 138 countries had ratified, acceded to, or approved the Convention.

The Convention provides a framework that enables governments to work together to achieve the ultimate objective of stabilizing GHG concentrations in the atmosphere at a level that would prevent dangerous anthropogenic interference with the climate system. The Convention recognizes and takes into consideration the special situations of different groups of countries.

Party to the Convention is expected to implement certain commitments, such as the following:

- Develop, update, and submit national inventories of anthropogenic emissions by sources and removals by sinks of GHGs that are not controlled under the Montreal Protocol on Substances that Deplete the Ozone Layer
- Formulate, implement, publish, and regularly update national programs for mitigating and adapting to climate change
- Promote and cooperate in the development, application, and diffusion of technologies and processes that will reduce the emissions of GHGs and conserve or enhance the sinks of GHGs
- Strengthen scientific and technical research and systematic observations related to climate systems to further understanding and to reduce or eliminate the remaining uncertainties about the causes, effects, magnitude, and timing of climate change and the economic and social consequences of various response strategies.

The Convention does not specify mandatory emission reductions that apply to all Parties. However, the commitment of the developed countries (that is, Annex I Parties) includes a target reduction of GHG emissions to 1990 levels by the year 2000.

The First Conference of the Parties to the Convention (COP–1) was held in Berlin from March 21, to April 7, 1995. Major decisions included the following:

- Establishment of a permanent secretariat located in Bonn and the adoption of the budget for the biennium 1996–1997
- Continuation of the Global Environmental Facility (GEF) as the interim financial mechanism, the arrangements between the COP and the financial mechanism, as well as the initial guidance on policies, program priorities, and eligibility criteria for the financial mechanism
- Establishment of the two subsidiary bodies—the Subsidiary Body for Scientific and Technical Advise (SBSTA) and the Subsidiary Body for Implementation (SBI)—to begin their work in October 1995;
- Agreement to start a pilot phase for activities implemented jointly by Annex I parties and non-Annex I parties on a voluntary basis to reduce the emissions of GHGs into the atmosphere
- Review of national communications from Annex I Parties and the first communications from parties not included in Annex I.

One of the important issues discussed at COP–1 was the adequacy of commitments. The majority of the Parties represented at COP–1 recognized that the existing commitments under the convention were inadequate and, hence, agreed upon the Berlin Mandate—a mandate for negotiation of a protocol. For the next two years, the focus of work for the parties will be to negotiate a protocol or other legal instrument that would enable the COP to take appropriate action for the period beyond the year 2000, including action to strengthen the commitments of Annex I parties in article 4, paragraphs 2(a) and (b) of the convention. The parties hope to achieve the widest possible cooperation by all countries, in accordance with their common but differentiated responsibilities, their respective capabilities, and their social and economic conditions. Negotiations will cover all GHGs, their emissions by sources, and removal by sinks in all relevant sectors. The process will call for Annex I parties to elaborate policies and measures and to set quantified objectives within a specified time, such as by 2005, 2010, and 2020, for reducing anthropogenic emissions and increasing removals by sinks of GHGs not controlled by the Montreal Protocol on Substances that Deplete the Ozone Layer.

3. The UNEP Atmosphere Sub-programme

UNEP contributes to the international work on climate change through its Atmosphere Sub-programme, which addresses three basic concerns: risks to the ozone layer, atmospheric pollution and its transport, and climate variability and change.

UNEP has been responsible for the WCIRP under the WCP. Many of UNEP's activities in the field of climate change fall within the framework of the WCIRP. Specific activities currently implemented by UNEP under the WCIRP include: initiating and supporting national and regional studies to assess the impacts of climate change and sea-level rise; providing countries with assistance in identifying response strategies to mitigate or adapt to climate variability and change; and organizing regional workshops on the use of El Niño Southern Oscillation information to predict droughts and floods and increase preparedness for such events.

3.1. Country Study Projects

UNEP's country study projects focus mainly on case studies, with the aim of advancing methodology development and establishing common guidelines or a common methodological framework for climate-related country studies. The projects are carried out in close collaboration with the IPCC to ensure that the UNEP projects complement IPCC efforts to provide assessments and scientific input for the implementation of the UNFCCC. The Atmosphere Sub-programme encompasses three country study projects, described in the following sections.

3.1.1. Sources and sinks of greenhouse gases

UNEP is supporting the development and testing of an international standard methodology for producing national inventories of the sources and sinks of GHGs. This effort is conducted through a GEF-funded project on "Country Case Studies on Sources and Sinks of Greenhouse Gases." Under the current phase of the project, direct technical support is provided to nine developing countries (Costa Rica, The Gambia, Mexico, Morocco, Poland, Senegal, Tanzania, Uganda, and Venezuela) to develop government-sanctioned, national GHG inventories. The inventories are developed using the IPCC Guidelines (IPCC/OECD, 1994; UNEP *et al.*, 1995), the only international standard methodology the Parties to the Convention have accepted for calculating and reporting national GHG inventories. Testing and application of the methodologies in these countries has provided valuable inputs for refining the IPCC guidelines.

As a part of this project, UNEP is working in collaboration with other international and bilateral country studies programs (including those of the United Nations Development Programme (UNDP), START, the United States, Germany, and The Netherlands) to support a series of regional workshops on GHG inventories in Africa, Asia, Latin America, and Eastern Europe. The workshops assemble national experts and technicians from approximately 100 countries as well as numerous nongovernmental organizations and oriented them to the methodology and its application. The UNEP project has also supported national training workshops in each of the nine study countries.

Under the project, an electronic mail network and computer conference was established in Africa by Climate Network Africa (CNA), working as a contractor in collaboration with GREENET. The network enabled the six African countries participating in the project to communicate with each other, with UNEP, and with several international agencies.

To continue supporting the UNFCCC's long-term objective to mitigate and adapt to the threat of climate change, UNEP has developed a Phase II project proposal to be considered for GEF funding. Phase II will contribute to the further development and refinement of the GHG inventory methodologies. The Phase II proposal builds upon the experience gained through implementation of Phase I, and takes into account the recommendations of country study teams, contractors, and collaborating institutions. The specific activities proposed in Phase II include the following:

- Further refining the IPCC GHG inventory methodology for universal application, including incorporating methods for additional GHGs; making adjustments to methods in land-use change and forestry; and supporting further research on nation/region/ecosystem-specific emission factors.
- Supporting a series of national studies in four countries.
- Harmonizing the guidelines with methodologies being developed for other climate-related issues.
- Promoting information exchange through electronic networks, extending the electronic network to other teams participating in country studies, and developing and increasing the availability of technical material on GHG inventory methodologies.

3.1.2. Methodologies for assessing climate impacts and adaptation strategies

Under the UNFCCC, parties to the convention are requested to formulate national programs containing measures to them to adapt to the adverse effects of climate change. COP–1 agreed on a three-stage approach towards adaptation. In the short term, referred to as Stage I, the primary activity envisaged is planning, which includes "studies of possible impacts of climate change to identify particularly vulnerable countries or regions and policy options for adaptation, and appropriate capacity building."

UNEP, in collaboration with the IPCC, other scientific institutions and international organizations, is responding to these needs by developing guidelines and methodologies for assessing the impacts of climate change, by developing adaptation strategies, and through testing the guidelines and methodologies under field conditions in country case studies. GEF has recently approved funding for these country studies, and detailed project preparations are under way. A handbook on methods for assessing climate impacts and adaptation will also be prepared by UNEP in cooperation with the IPCC and other relevant institutions.

This project will produce recommendations on methodologies for studying impacts and adaptation options, especially the advantages and limitations of different methods and approaches. It will also develop a set of illustrative country case studies, which will illuminate possible other adaptation options.

3.1.3. Greenhouse gas abatement costing studies

All parties to the UNFCCC have made a commitment to develop national plans containing measures that would mitigate climate change by addressing all sources and sinks of GHGs. The convention also states that mitigation should be cost-effective.

In 1991, UNEP initiated a project entitled, "Preparation of a Methodology to Undertake National Greenhouse Gas Abatement Costing Studies." The objective of the project was to develop a methodological framework for determining costs of reducing GHG emission reductions at country level and to test this frameworks in a broad set of representative countries. During the project's first phase, completed in 1992, the subject was reviewed and an attempt was made to formulate an approach for comparable national costing studies. The second phase of the project aimed at producing a detailed set of methodological guidelines and testing and refining these through country studies in ten countries—Brazil Denmark, Egypt, France, India, the Netherlands, Senegal, Thailand, Venezuela, and Zimbabwe. During this phase, the methodological guidelines were formulated and continually revised in conjunction with the national teams, particularly through four project workshops. The third phase of the project involved detailed studies in Venezuela and Zimbabwe, as a result of which the methodological guidelines were expanded to better address nonenergy and non-carbon dioxide mitigation options.

A new project entitled, "Economics of GHG Limitations—Phase I: Methodological Framework for Climate Change Mitigation Assessment" has been approved for funding under GEF. The goal of this project, which builds on the initial UNEP project, is to further improve the methodological framework for calculating the costs of mitigation options and to establish a framework for integrated assessment of mitigation strategies. Eight country case studies are planned.

This methodology will be an input for the work of the IPCC and the UNFCCC implementation process. In particular, it will enable the parties to develop their national mitigation strategies and analyses as well as to prepare national communications concerning their implementation of the Convention. One important part of this project is to help developing countries build capacity for conducting national GHG abatement analyses. The UNEP Collaborating Center on Energy and Environment (UCCEE) at RISO National Laboratory, Denmark, is responsible for the project. It will be implemented in cooperation with Lawrence Berkeley National Laboratory (LBNL).

3.2. Information Exchange and Public Awareness

Exchanging information and raising public awareness are also important activities under the Atmosphere Sub-programme. In response to the need for information on climate-related country activities, UNEP, together with the UNFCCC Interim Secretariat created a joint project in 1993 entitled, "Country Activities on Climate Change: Information Exchange System (CC:INFO)." The main objective of the project is to provide up-to-date information on who is doing what in the field of climate change and where the available technical and financial resources are. This project makes it easier to match demand for the resources needed to implement climate related activities with available financial and technical resources. The information collected under CC:INFO is stored in a computerized database and disseminated in the form of reports consisting of country and organization profiles, computer diskettes, and on-line system (Internet).

In addition, detailed information dissemination and public awareness work has been conducted by UNEP's Information Unit on Climate Change (IUCC), which was established in 1991 to provide policymakers and opinion leaders with timely, accurate, and action-oriented information on all aspects of climate change. In carrying out this mandate, IUCC has helped to raise international awareness of this critical issue by producing and disseminating information products such as a dossier of factsheets covering the causes of climate change, likely impacts, and possible response strategies; videos on climate change and video catalogues; Climate Change Bulletins; and a "Convention Kit" containing the Climate Change Convention and an explanatory leaflet. IUCC also provides information services including organizing public awareness seminars, arranging media campaigns for the UNFCCC, and providing conference support.

3.3. Networking and Coordination

During 1994 and 1995, the Atmosphere Sub-programme has concentrated on strengthening coordination of the climate impact and response strategy activities rather than on conducting a research-based program. As a part of the coordination mechanism, UNEP is establishing and operating regional Climate Impacts and Response Strategies Networks (CIRSNet) starting with a regional network for Africa. CIRSNet/Africa provides participants with a forum for exchanging information and experiences and for discussing various problems related to climate change and variability so they can identify actions that are needed in the region and in the individual countries of Africa.

Currently, twenty countries in Africa are participating. Each country has designated a national focal point for CIRSNet/Africa. The First Workshop of CIRSNet/Africa was held from July 11 to 14, 1995 in Niamey, Niger. The Workshop produced a set of recommendations on what actions can be taken as initial steps towards starting or strengthening climate-related activities in the region.

Under CIRSNet/Africa, UNEP is working with United Nations bodies including the UNFCCC Secretariat, UNDP, United Nations Institute for Training and Research (UNITAR), and WMO, as well as with regional organizations such as the African Center for Meteorological Application for Development (ACMAD), and Environmental Development Action in the Third World (ENDA–TM) to implement the activities under the network. This includes working closely with the national focal points to establish or strengthen national institutional frameworks and to develop and implement national climate-related activities. The various actions that the countries decide to take can then be formulated into projects as part of national programs. Priorities can be established and funding sought for these programs. UNEP hopes that these efforts will lead to a strengthening of regional and national capacity in developing countries to address climate variability and change and to implement the UNFCCC and the WCP. CIRSNet/Africa uses the CC:INFO information base to exchange information with international and other organizations and with countries in other regions.

In addition to the CIRSNet activities, UNEP is seeking to mobilize expertise from the social and economic sectors to assist nations in developing mitigation and response strategies. Towards this end, UNEP proposes to work closely with the Human Dimensions Programme (HDP) and associated institutions. If funds become available, UNEP will pursue a program that would enhance cooperation between social and natural scientists in their understanding climate change, its impacts, and possible policy options.

4. Conclusions

Considerable scientific uncertainty remains with regard to the extent, magnitude, and rate of climate change resulting from human activities and the environmental and the socioeconomic impacts of such change. The natural variability of climate makes assessing the human-induced climate change difficult. Although the magnitude of global average warming from increased levels of GHGs in the atmosphere has been predicted, how it will affect the climate and weather on the regional, national, and local scale is unknown. Furthermore, impacts of climate change on the environment and the society are complex interactions.

While experts continue their efforts to develop scientific knowledge on climate change, actions to address climate change are underway both internationally and nationally. But further actions are needed to eliminate or minimize the risk of severe, adverse impacts on the environment and on society.

UNEP has always advocated a cautious approach in dealing with scientific uncertainties. It will continue to catalyze and initiate actions on climate change and encourage and to help decisionmakers to identify and implement response strategies so society can continue to develop in a sustainable manner, with full consideration of the needs of future generations that will inhabit this earth.

References

World Meteorological Organization (WMO): 1979, Proceedings of the World Climate Conference, WMO #537, Geneva, February.

WMO: 1993, *The Climate Agenda,* Intergovernmental Meeting on the World Climate Programme, Meeting Statement and Report, Geneva, April 14–16.

UNEP/Atmosphere: 1995, *Report of the Meeting of the Eleventh Session of the Scientific Advisory Committee of the World Climate Impacts and Response Strategies Programme, December 12–14, 1994, Amsterdam, The Netherlands,* Atmosphere Report Series no. 1.

FAO, ICSU, UNEP, UNESCO and its IOC, and WMO: 1995, *The Climate Agenda, International Climate-Related Programmes: A Proposal for an Integrating Framework.*

IPCC: 1990, *Climate Change: The IPCC Scientific Assessment,* Houghton, J.T., Jenkins, G.J., and Ephraums, J.J. (eds)., Cambridge, U.K.: Cambridge University Press.

IPCC: 1990, *Climate Change: The IPCC Impacts Assessment,* Tegart, W.J. McG., Sheldon, G.W., and Griffith, D.G. (eds.), Canberra: Australian Government Publishing Service.

IPCC: 1990, *Climate Change: The IPCC Response Strategies,* Response Strategies Working Group III, UNEP/WMO.

IPCC: 1992, *Climate Change 1992: The Supplementary Report to the IPCC Scientific Assessment,* Houghton, J.T., Callander, B.A., and Varney, S.K. (eds.), Cambridge U.K.: Cambridge University Press.

IPCC: 1992, *The IPCC 1990 and 1992 Assessments—IPCC First Assessment Report Overview and Policymaker Summaries and 1992 IPCC Supplement.*

IPCC: 1992, *Climate Change 1992: The Supplementary Report to the IPCC Impacts Assessment,* Tegart, W.J. McG., and Sheldon, G.W. (eds.), Canberra: Australian Government Publishing Service.

IPCC: 1994, *Climate Change 1994: Radiative Forcing of Climate Change and An Evaluation of the IPCC IS92 Emission Scenarios,* Houghton, J.T., Meira Filho, L.G., Bruce, J., Lee, H., Callander, B.A., Haites, E., Harris, N., and Maskell, K. (eds), Cambridge, U.K.: Cambridge University Press.

IPCC: 1994, *Radiative Forcing of Climate Change: The 1994 Report of the Scientific Assessment Working Group of IPCC, Summary for Policy Makers,* UNEP/WMO.

IPCC: 1994, *IPCC Technical Guidelines for Assessing Impacts of Climate Change Impacts and Adaptations,* Working Group II of the IPCC, University College London, London, U.K. and National Institute for Environmental Studies, Tsukuba, Japan.

United Nations: 1992, *United Nations Framework Convention on Climate Change.*

IPCC/OECD (Intergovernmental Panel on Climate Change/Organisation for Economic Co-operation and Development) Joint Programme: 1994, *IPCC Draft Guidelines for National Greenhouse Gas Inventories,* IPCC/OECD Joint Programme, Paris, 3 volumes.

UNEP, OECD, IEA, and IPCC (United National Environment Programme, Organisation for Economic Co-operation and Development, International Energy Agency, Intergovernmental Panel on Climate Change): 1995, *IPCC Guidelines for National Greenhouse Gas Inventories,* IPCC, Bracknell, 3 Volumes.

PART III

GREENHOUSE GAS EMISSION INVENTORIES AND MITIGATION OPTIONS: METHODOLOGICAL ISSUES

USE OF FOREST INVENTORIES AND GEOGRAPHIC INFORMATION SYSTEMS TO ESTIMATE BIOMASS DENSITY OF TROPICAL FORESTS: APPLICATION TO TROPICAL AFRICA

S. BROWN and G. GASTON

US Environmental Protection Agency, 200 SW 35th St., Corvallis, OR 97333, USA

Abstract. One of the most important databases needed for estimating emissions of carbon dioxide resulting from changes in the cover, use, and management of tropical forests is the total quantity of biomass per unit area, referred to as biomass density. Forest inventories have been shown to be valuable sources of data for estimating biomass density, but inventories for the tropics are few in number and their quality is poor. This lack of reliable data has been overcome by use of a promising approach that produces geographically referenced estimates by modeling in a geographic information system (GIS). This approach has been used to produce geographically referenced, spatial distributions of potential and actual (circa 1980) aboveground biomass density of all forests types in tropical Africa. Potential and actual biomass density estimates ranged from 33 to 412 Mg ha^{-1} (10^6g ha^{-1}) and 20 to 299 Mg ha^{-1}, respectively, for very dry lowland to moist lowland forests and from 78 to 197 Mg ha^{-1} and 37 to 105 Mg ha^{-1}, respectively, for montane-seasonal to montane-moist forests. Of the 37 countries included in this study, more than half (51%) contained forests that had less than 60% of their potential biomass. Actual biomass density for forest vegetation was lowest in Botswana, Niger, Somalia, and Zimbabwe (about 10 to 15 Mg ha^{-1}). Highest estimates for actual biomass density were found in Congo, Equatorial Guinea, Gabon, and Liberia (305 to 344 Mg ha^{-1}). Results from this research effort can contribute to reducing uncertainty in the inventory of country-level emission by providing consistent estimates of biomass density at subnational scales that can be used with other similarly scaled databases on change in land cover and use.

1. Introduction

Concern over increasing amounts of greenhouse gases in the atmosphere and their potential to influence global climate change has produced a number of initiatives, including the United Nations Framework Convention on Climate Change. Country-level activities under this framework agreement include estimations of greenhouse gas emissions from all sources within that country. Carbon dioxide (CO_2) is an important greenhouse gas, and while fossil fuel burning is a major source of atmospheric CO_2, change in land cover and land use can be an important source, especially in the tropics (Dixon *et al.*, 1994).

The quantity of biomass carbon (C) in a given ecosystem is one of the most uncertain factors involved in estimating changes in C flux from terrestrial ecosystems (Brown *et al.*, 1989). While land-cover changes can be seen with remote sensing data, biomass-carbon density and actual land-use changes can only be inferred. Very few studies have actually measured the biomass in tropical forest ecosystems (Brown and Lugo, 1982), and although it is possible to estimate biomass from a complete forest inventory (Brown *et al.*, 1989), the lack of sufficient, reliable inventory data is a significant factor in limiting biomass estimates for tropical forests.

Environmental Monitoring and Assessment **38**: 157-168, 1995.

What is needed is a method that can extend the few reliable biomass-carbon density estimates to larger, national and regional scales. Brown *et al.* (1993) and Iverson *et al.* (1994) developed such a method for forests in tropical Asia. This modeling approach used a geographic information system (GIS). The results of this work demonstrated that it is feasible to produce a spatial distribution of biomass densities for tropical forests.

The purpose of this paper is to report on efforts to develop a spatial distribution of aboveground biomass density of woody vegetation in tropical Africa using the methodology described by Brown *et al.* (1993) and Iverson *et al.* (1994). This methodology was used to estimate biomass-carbon density for closed forests, open forests, woodlands, and savannas. Aboveground biomass density is reported by country and by ecological zone. This paper also shows how the approach can be used to provide estimates of biomass density for subnational units within the larger countries of tropical Africa.

2. Methods

2.1. Study Area

The African land mass is primarily tropical, with a wide variety of vegetation communities. Deserts bracket extensive areas of savanna and woodland, which are the characteristic vegetation communities of Africa. The central core of the continent (the Zaire River Basin and Western Africa) contains large areas of moist tropical forests. Variations in the distribution of potential ecological communities depend primarily on climatic factors, topography, and soils. The actual distribution of vegetation communities is very much related to density and type of human activity. Most human activity is concentrated in the seasonal forests, woodlands, and savanna ecosystems. The study area for this research effort included those countries that for the most part are located in tropical Africa. Nontropical Southern Africa (South Africa, Lesotho, and Swaziland) and Mediterranean Africa (Egypt, Libya, Algeria, Morocco, and Western Sahara) were excluded.

2.2. Geographic Information System Tools and Data

Spatial analysis, modeling, and overlay of spatial data were important tools in this research effort. This section briefly describes the spatial data and tools used. The GIS software used was ARC/INFO, with a heavy reliance on the raster overlay and modeling capabilities of the GRID module. The projection was a rectangular equal-area projection, with a pixel size of 5000 m by 5000 m.

Spatial databases for Africa came from the following of sources:

1. Population density for 1960, 1970, 1980, and 1990 for nations and subnational units (the area of a subnational unit is generally larger than the area of a pixel) were provided by the Food and Agriculture Organization's (FAO) Tropical Forest Resource Assessment 1990 Program (FRA90) (M. Lorenzini and K.D. Singh, 1995).
2. Elevation data were taken from a global digital elevation database (referred to as ETOPO5), which originated at the U.S. National Geophysical Data Center. The elevation

data were rescaled into 15 m increments and all bathymetry data were calculated to zero.

3. Soils data came from the FAO Soils Map of the World (FAO, 1971–81). This map was reclassified into simplified maps of five soil texture classes, five slope classes, and two soil depth classes.
4. The FAO agrometeorological database was used to develop climate surfaces. Various climatic parameters such as mean monthly precipitation, maximum temperature, minimum temperature, day-night temperature differential, evapotranspiration, and vapor pressure were recorded for each station. Using the spatial coordinates for each station, it was possible to create two interpolated surfaces: annual precipitation and an integrated climatic index, which incorporated precipitation and length of the growing season (Iverson *et al.*, 1994).
5. Distribution of vegetation was obtained from a vegetation map, circa 1980, which was produced by the FAO in collaboration with the International Institute of Vegetation Mapping, Toulouse, France (Lavenu, 1987). The original vegetation classes were simplified to produce a map showing closed forest, woodland and wooded savanna, grassland savanna, and other classes, which included agriculture. A map of ecofloristic zones (EFZ), developed for the FAO FRA90 program (Sharma, 1986), was used with various reclassification options as a verification tool and for summarizing results. Sixty-four EFZs were distinguished for Africa based on floristic composition and climatic, edaphic, and physiographic factors.

2.3. Modeling Potential Forest Biomass

The current distribution of forest biomass is assumed to be a combination of the potential biomass density, which is based on prevailing climatic, edaphic, and geomorphological conditions; natural disturbances; and the cumulative impact of human activities that reduce biomass. For the African forests, natural disturbances such as tropical cyclones or storms were assumed to be minimal. Wildfires in woodlands and wooded savannas do occur and are assumed to be part of the human disturbance regime (Goldammer, 1993). They are implicitly dealt with later in the analysis. The model used to estimate potential biomass (described in detail in Brown *et al.*, 1993, and Iverson *et al.*, 1994) performed a weighted overlay of input layers: precipitation, climatic index, elevation and slope, and soil texture. Each of the four input layers was assigned a maximum score of 25 points and a scaled relationship depicting how biomass is affected by a given layer alone. Initial weighting factors were based on published information, and results were compared to values at known localities and were evaluated by colleagues familiar with the region. Initial weightings were adjusted through an iterative process. The final iteration produced a raster grid for tropical Africa with each pixel containing a potential biomass density (PBD) index ranging in value from about 40 to 100.

The above methodology is designed to generate biomass density estimates for *woody* formations only, so the next step was to remove nonwoody formations from the map of potential biomass density indices. To accomplish this, the FAO vegetation map was reclassified (Lavenu, 1987). All woody formations, including all classes of disturbed forests, woodlands, and mosaics, were classed into either closed forests or woodland/wooded savanna; all nonwoody formations were removed. The map of PBD indices for tropical Africa was then overlain with this map to remove nonwoody formations.

Next, the PBD indices were converted into biomass density. This required the assignment of biomass density estimates across the range of index values. The most critical values are those that identify the upper and lower biomass limits. A limited set of ecological studies that provided biomass estimates for mature forests, woodlands, and wooded savannas was used to establish the upper and lower limits of biomass density (Brown and Lugo, 1982; estimates from data in Pierlot, 1966; Iverson *et al.*, 1994).

2.4. The Degradation Ratio

In any system, a variety of natural and anthropogenic factors reduce biomass from its potential. Long-term human use such as fuelwood gathering, logging (both sanctioned and unsanctioned), agricultural clearing, grazing, shifting of cultivation, and anthropogenic burning all reduce the amount of biomass present (Brown *et al.*, 1991). Indeed, in Africa long-term human use has produced a suite of ecosystems that may no longer contain a "natural" (non-human-impacted) ecosystem.

Population density has been identified as an index that can be used to quantify the long-term impact of human actions on the biomass density of forests in tropical Asia (cf. Brown *et al.*, 1993). While there is some disagreement about the relationship of population density and biomass reduction, in most areas of tropical Asia where the population lives in close contact with forests and other wooded areas there seems to be a significant, direct relationship between population density and a reduction in the potential biomass density. The percentage of forest cover in tropical regions has also been shown to correlate closely with population density at subnational levels when stratified into ecological zones (FAO, 1993). Although population density seems to work well as an explanation for trends in biomass and forest area reduction, it does not account for all of the variability. Other socioeconomic factors such as the state of industrialization, the ratio of rural to urban population, the importance of subsistence use of land, and the export or import of timber and agricultural products also contribute to a reduction in biomass and forest cover (Flint and Richards, 1994). However, continued increases in population density do not necessarily cause continued decreases in biomass and forest cover. Eventually a population density is reached beyond which there is no further decrease in forest cover and biomass change. At this point other factors come into play, and forest cover and biomass density may even increase with increasing population density, such as is occurring in many of the industrialized countries of the temperate zone. However, for most tropical countries this limit appears not to have been reached.

The amount of biomass reduction (degradation) has been estimated by comparing biomass density estimates from forest inventories conducted at the subnational to national scale with the modeled estimate of potential biomass density for the same inventory location (Brown *et al.*, 1993; Iverson *et al.*, 1994). This comparison produced "degradation ratios," unitless factors that relate the reduction of potential biomass density to the actual biomass density (see Figure 1). Data on the corresponding population density

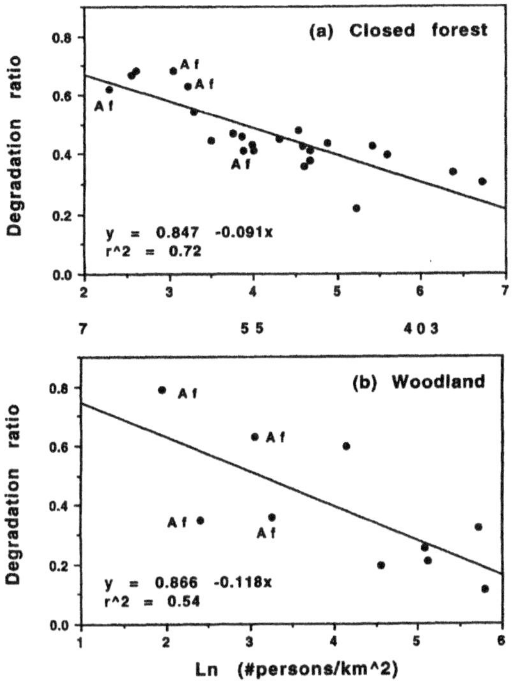

Figure 1. Relationship between degradation ratio and natural log of population density for (a) closed forests (n = 25) and (b) woodlands/savannas (n = 10).

for the decade of the inventory was also obtained from the population data set (see Section 2.2). This database was then stratified into climatic zones (for example, moist or dry).

Based on twenty-seven inventory units, Brown *et al.* (1993) compiled a database of degradation ratios and corresponding population densities for moist forests and dry forests of tropical Asia. We identified eight reliable inventory-based data sets for estimating actual biomass density in tropical Africa. Four were located in the moist closed forest zone: Gabon (Clement and Nouvellet, 1978), northern and southern Cote d'Ivoire (Development and Resources Corporation, 1967; Republique de Cote d'Ivoire, 1975), and Sierra Leone (United Nations Development Program, 1972), and four in the dry woodland/wooded savanna zone of Cameroon, Mali, Tanzania, and Zambia (FAO, 1989). Using a variety of methods described in Brown *et al.* (1989), Brown and Iverson (1992), and Brown and Lugo (1992), we estimated the actual biomass density from these African inventories.

The data for the closed forest zone of Africa were combined with the moist forest data from Asia. The African data appeared to fit well with the Asian database, and combining the data slightly improved the fit of a linear regression line of degradation ratio versus natural log (ln) of population density (from adjusted $r^2 = 0.70$ for Asia alone to adjusted $r^2 = 0.72$ for Asia and Africa; Figure 1). Similarly, the data for the dry woodland/savanna in Africa were combined with the data for the dry Asian forests. Again, the overall fit of the points was good and improved the adjusted r^2 of the linear regression (from 0.41 for Asia alone to 0.54 for Asia and Africa; Figure 1).

2.5. Actual Biomass Density Map

The actual biomass map was the product of the potential biomass density map and the distribution of degradation ratios. The population density map at the subnational level for 1980 was stratified by forest type (closed forest and woodland/wooded savanna) and a degradation ratio calculated for each pixel by using the regression equations in Figure 1. While each pixel of a subnational unit has the same population density value across the unit, the forest strata may vary. Thus, the calculated degradation ratio could also vary within a unit. The estimates of actual biomass density were calculated on a pixel-by-pixel basis, with the value in each pixel reflecting the predicted estimate of potential biomass density, the forest-type stratification, and the population density for 1980 from each subnational unit.

3. Results and discussion

Because forest inventories in tropical Africa are limited and because most of these were used to calibrate the model, regionwide verification of the results was difficult. The few inventories that were not used in the calibration gave the following comparisons (ongoing research by authors):

- Guinea—the model gave a national weighted mean biomass density of 140 Mg ha^{-1}; and the inventory, a value of 135 Mg ha^{-1}.
- Gambia—the model gave a national weighted mean biomass density of 29 Mg ha^{-1}; and the inventory, a value of 36 Mg ha^{-1}.
- Mozambique—the model gave a national mean weighted biomass density of 55 Mg ha^{-1}; and the inventory, a value of 57 Mg ha^{-1}.
- Burkina Faso—the model gave a regional weighted mean biomass density for the tree savannas of 34 Mg ha^{-1}; and the inventory, a value of 21 Mg ha^{-1}.

Although only a few countries are compared here, the comparisons are in the right direction, with most model results practically the same as the inventory results.

To see if the results of the modeling seemed reasonable and yielded expected trends across the continent, the biomass density estimates were aggregated by EFZ. The EFZ map was reclassified to show four lowland zones and two montane zones akin to the Holdridge life zone system. This map served as an independent check because it had not been used in any stage of the analysis. Estimated potential biomass densities were summarized for each zone (Table I). As expected, the estimates are highest in the lowland moist zone (412 Mg ha^{-1}) and decrease with increasing aridity. The lowest biomass density was obtained for the lowland very dry zone (33 Mg ha^{-1}). Montane zones have slightly

TABLE I

Mean potential and actual aboveground biomass density (Mg ha^{-1}), coefficients of variation (CV in %), and degradation ratios (DR) for tropical forests of Africa by ecofloristic zone.

Ecofloristic	Potential	CV	Actual	CV	DR
Zone					
Lowland moist	412	15	299	22	0.73
Lowland seasonal	211	60	141	71	0.67
Lowland dry	92	96	60	112	0.65
Lowland very dry	33	112	20	110	0.61
Montane moist	197	55	105	61	0.53
Montane seasonal	78	123	37	111	0.47

Note: Aboveground biomass estimates are based on 1980 population densities.

less biomass than their lowland counterparts, as previously observed by Brown and Lugo (1982). These expected patterns of potential biomass density by EFZ are convincing evidence that the methodology used for generating a biomass density map of Africa produces reasonable estimates at a continental scale, as was the case for tropical Asia (Brown *et al.*, 1993).

The variability in potential biomass density, as expressed by the coefficient of variation (CV), is lowest in the lowland moist zone (15%), where moisture is generally not limited, and one can reasonably expect a relatively homogenous landscape with small variations in biomass. As the climatic zones become arid, the heterogeneity of potential biomass density appears to increase across the landscape (high CVs), most likely the result of the differential effects of soils and slope on biomass in the fairly uniform dry environment.

3.1. Distribution of Potential and Actual Biomass Density Estimates

Examination of the potential biomass density map (Figure 2) shows that the overall distribution of biomass appears to match the expected patterns. High biomass densities (>300 Mg ha^{-1}) are found in the moist tropical forest zones of central (Zaire Basin) and west Africa. Moving away from these zones, biomass density decreases where woodland/savanna zones tend to dominate the landscape. Woody formations with the lowest potential biomass densities are located in the eastern and southern parts of tropical Africa and in countries of the Sahel.

The distribution of the biomass density of tropical African forests changes dramatically under the influence of human activities (Figure 3). Practically all forest areas with potential biomass density greater than 400 Mg ha^{-1} have disappeared in the map of actual biomass density (Figure 3). Furthermore, in the actual biomass density map, the

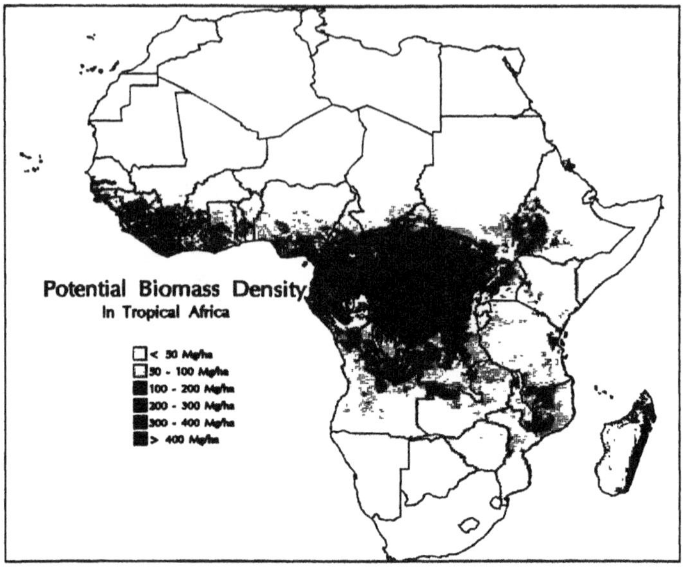

Figure 2. Map of potential biomass density for woody formations of tropical Africa. The original map has 12 classes.
Note: A copy can be obtained from this paper's authors.

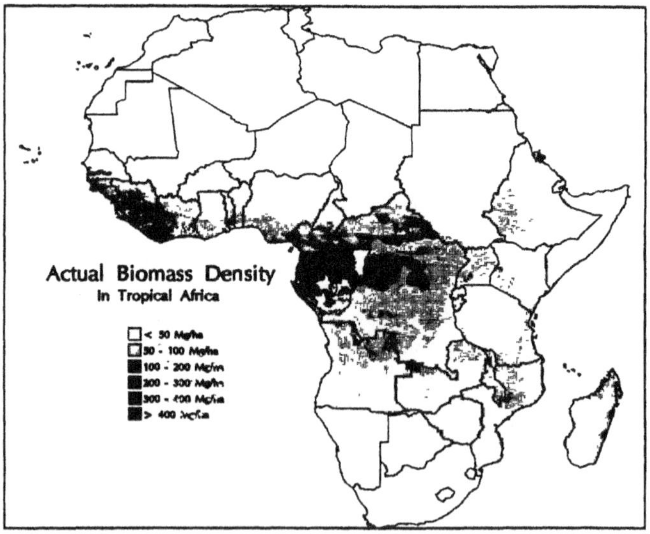

Figure 3. Map of actual biomass density for woody formations of tropical Africa. The original map has 10 classes (because of degradation, the highest classes of the potential biomass density map are missing).
Note: A copy can be obtained from this paper's authors.

distribution of the three biomass density classes greater than 200 Mg ha^{-1} has become severely limited.

Actual biomass density estimates range from about 300 Mg ha^{-1} in the lowland moist zone to 20 Mg ha^{-1} in the lowland very dry zone (Table I). Forest biomass was reduced or degraded more from its potential in montane zones (47%–53%) than in lowland zones (61%–73%). Adding human influence to the analysis, as reflected by the estimates of actual biomass density, results in a higher CV for the lowland moist, seasonal, and dry zones and for the montane moist zones, from a range of 15% to 96% to a range of 22% to 112% (Table I). The CVs for actual biomass density for the lowland very dry and montane seasonal zones decreased slightly (112% to 110% and 123% to 111%) (Table I).

3.2. Biomass Density Estimates at the Country Level

Biomass density estimates can be extracted on a country-by-country basis (Table II). The countries with the highest estimates of potential biomass density were Liberia (466 Mg ha^{-1}), Equatorial Guinea (442 Mg ha^{-1}), Sierra Leone (411 Mg ha^{-1}), Gabon (375 Mg ha^{-1}), and Congo (374 Mg ha^{-1}). These countries also had low CVs and are characterized as small to medium in size and dominated by the lowland moist climatic zone. The countries with the lowest potential biomass density estimates were Botswana (15 Mg ha^{-1}), Niger (16 Mg ha^{-1}), and Somalia (20 Mg ha^{-1}), all large countries dominated by the lowland very dry climatic zone.

Of the thirty-seven countries included in this study, more than half contained forests that had less than 60% of their potential biomass. The most degraded countries were Senegal (33% of potential), Burundi (36% of potential), and Rwanda (38% of potential). The least degraded countries were Congo (92%), Botswana (91%), and Gabon (90%). Spatial heterogeneity within a country, as reflected in CVs, appears to be related to its size, population density, and domination by dry climatic zones. High population and large size both favor high variability in biomass density estimates in dry countries such as Ethiopia, Sudan, and Kenya (Table II).

In many cases, the country level is the minimum spatial resolution at which biomass density estimates can be made with any degree of confidence because of the coarse resolution of several of the data layers (for example, FAO soil maps and elevation maps) used in making the map of potential biomass density. However, for some of the larger countries with large subnational units, biomass density estimates can be reported by subnational unit with some confidence. For example, Zaire, which is one of the largest countries in tropical Africa, is divided into nine major subnational units with corresponding biomass density estimates (Table III). The estimates for potential and actual biomass density are highest (424 Mg ha^{-1} and 321 Mg ha^{-1}) with the lowest CVs (16% and 17%) in Equateur, which, as the name suggests, is the province in Zaire that is along the equator and is dominated by the Zaire River basin. The lowest potential and actual biomass density estimates (146 Mg ha^{-1} and 100 Mg ha^{-1}) and highest CVs (45% and 56%) are in Shaba Province, which is farthest south and farthest from the core area of moist tropical forest. The effect of high population pressure on biomass estimates can be clearly seen in Kinshasa, where the high population density (215 persons km^{-2}) has reduced the potential biomass density by about two-thirds.

TABLE II

Area-weighted mean potential and actual aboveground biomass density (Mg ha^{-1}), coefficient of variation (CV in %), and the degradation ratio (DR, ratio of actual to potential biomass density) for tropical African forests, by country

Country	Potential	CV	Actual	CV	DR
Angola	100	83	73	81	0.73
Benin	112	59	58	53	0.52
Botswana	15	56	13	55	0.91
Burkina Faso	65	66	34	70	0.53
Burundi	119	48	43	49	0.36
Cameroon	307	46	217	54	0.71
CAR	243	43	200	44	0.82
Chad	63	57	43	58	0.68
Congo	374	21	344	29	0.92
Cote d'Ivoire	276	41	165	44	0.60
Equatorial Guinea	442	9	318	10	0.72
Ethiopia	101	99	52	100	0.51
Gabon	375	18	339	19	0.90
Gambia	64	62	29	55	0.45
Ghana	182	51	83	57	0.45
Guinea	259	56	140	58	0.54
Guinea Bissau	153	46	85	47	0.55
Kenya	58	104	33	80	0.57
Liberia	466	10	305	22	0.65
Madagascar	322	39	196	37	0.61
Malawi	108	64	47	65	0.44
Mali	75	65	45	57	0.60
Mozambique	96	77	57	75	0.60
Niger	16	49	9	50	0.53
Nigeria	128	87	49	88	0.38
Rwanda	103	47	34	40	0.33
Senegal	50	88	32	75	0.62
Sierra Leone	411	21	199	26	0.48
Somalia	20	39	13	40	0.63
Sudan	95	93	64	95	0.67
Tanzania	83	75	45	69	0.55
Togo	155	57	72	58	0.46
Uganda	237	43	102	43	0.43
Zaire	297	42	206	47	0.69
Zambia	67	76	47	85	0.70
Zimbabwe	26	73	14	71	0.51

Note: Aboveground biomass estimates are based on 1980 population densities.

TABLE III

Mean potential and actual aboveground biomass density of woody formations (Mg ha^{-1}), coefficients of variation (CV in %), and population density (PD in #/km^2) for subnational units of Zaire

Subnational	Potential	CV	Actual	CV	PD
Bas-Zaire	193	33	98	36	33
Kinshasa	161	29	49	43	215
Shaba	146	45	100	56	7
Bandundu	248	32	157	35	11
Kasai-West	280	29	169	29	18
Kasai-East	288	27	176	28	13
Equateur	424	16	321	17	8
Kivu	292	31	156	46	7
Haut-Zaire	378	21	277	26	8

4. Conclusion

Reliable estimates of biomass density for tropical forests are important variables in assessing terrestrial carbon fluxes, but because of their limited availability, it has been necessary to develop alternative methods of extrapolating biomass density across the landscape at regional and national scales. The methods described in this paper make use of simple and widely available spatial data sets to extend limited biomass data to regional and national estimates. It appears that it is possible to use the biomass density estimates from ecological studies and biomass density estimates derived from forest inventory data to produce reasonable patterns in the distribution of biomass density in tropical forests. The patterns of biomass density distribution in all cases are consistent with expected patterns. Using population density as an index to model anthropogenic effects on biomass density appears to be valid statistically and may allow for projections to be made both forward and backward over a moderate amount of time (one to two decades). The ability to project biomass based on population trends, within the social and economic limits previously mentioned, is a powerful tool in assessing changes in carbon flux from the tropical forests.

Acknowledgments

This research was partially supported by a grant from the US Department of Energy (DOE DEFG02-90ER61081) to the University of Illinois and the US Environmental Protection Agency through its postdoctoral fellowship program, which is administered by the National

Research Council. The authors wish to thank A. Prasad for producing many of the GIS data layers used in the model, Alberto del Lungho for calculating some of the biomass estimates from forest inventory data, and Jack Winjum for helpful comments on the manuscript.

References

Brown, S. and Iverson, L. R.: 1992, Biomass estimates for tropical forests, *World Resource Review* **4**, 366-384.

Brown, S. and Lugo, A. E.: 1982, Storage and production of organic matter in tropical forests and their role in the global carbon cycle, *Biotropica* **14**, 161-187.

Brown, S. and Lugo, A. E.: 1992, Aboveground biomass estimates for tropical moist forests of the Brazilian Amazon, *Interciencia* **17**, 8-18.

Brown, S., Gillespie, A. J. R., and Lugo, A. E.: 1989, Biomass estimation methods for tropical forests with application to forest inventory data, *Forest Science* **35**, 881-902.

Brown, S., Gillespie, A. J. R. and Lugo, A. E.: 1991, Biomass of tropical forests of South and Southeast Asia, *Canadian Journal of Forest Research* **21**, 111-117.

Brown, S., Iverson, L. R., Prasad, A., and Liu, S.: 1993, Geographical distributions of carbon in biomass and soils of tropical Asian forests, *Geocarto International* **4**, 45-59.

Clement, J. and Nouvellet, Y.: 1978, *Inventaire Forestier dans la Zone d'Edenzork*, Republique du Gabon, Societe Gabonaise de Cellulose, Centre Technique Forestier Tropical, Nogent-sur-Marne, France.

Development and Resources Corporation: 1967, *Forestry Resources of the Southwest Region*, a report to the government of the Republic of Ivory Coast.

Dixon, R. K., Brown, S., Houghton, R. A, Solomon, A., and Wisnieski, J.: 1994, Carbon pools and flux of global forest ecosystems, *Science* **263**, 185-190.

Flint, E. P. and Richards, J. F.: 1994, Trends in carbon content of vegetation in South and Southeast Asia associated with changes in land use, in Dale, V. (ed.), *Effects of Land Use Change on Atmospheric CO Concentrations: Southeast Asia as a Case Study*, New York: Springer-Verlag, 201–300.

Food and Agriculture Organization, (FAO): 1971–81, *Soils Map of the World 1:5,000,000*, FAO-UNESCO.

FAO: 1989, *Studies on the Volume and Yield of Tropical Forest Stands 1. Dry Forest Formations*, FAO Forestry Paper 51/1.

FAO: 1993, *Forest Resources Assessment 1990 Tropical Countries*, FAO Forestry Paper 112.

Goldammer, J. G.: 1993, Historical biogeography of fire: tropical and subtropical, in Crutzen, P. J. and Goldammer, J. G. (eds.), *Fire in the Environment: the Ecological, Atmospheric, and Climatic Importance of Vegetation Fires*, Chichester, U.K.: John Wiley and Sons, 297–314.

Iverson, L. R., Brown, S., Prasad, A., Mitasova, H., Gillespie, A. J. R., and Lugo, A. E.: 1994, Use of GIS for estimating potential and actual biomass for continental South and Southeast Asia, in Dale, V. (ed.), *Effects of Land Use Change on Atmospheric CO_2 Concentrations: Southeast Asia as a Case Study*, Springer-Verlag, New York: ch. 3.

Lavenu, F.:1987, *Vegetation Map of Africa Scale 1:1,5000,000*, Institute de la Carte Internationale de la Vegetation, Universite Paul Sabatier, Toulouse, France, Distributed by FAO Forest Resources Assessment 1990 Project, Rome, Italy.

Pierlot, R.: 1966, Structure et composition de forets denses d'Afrique, especialement celles du Kivu, *Academie Royale des Sciences d'Outre-mer, Classes des Sciences Naturelles et Medicales*, N.S.–XVI–4, Bruxelles, Belgium.

Republique de Cote D'Ivoire, Ministere des Eaux et Foret: 1975, *Analyse et Commentaires des Resultats de l'Inventaire Forestier de la Region Nord-Ouest. Resultats de la Region du Nord-Ouest*, Inventaire Forestier National, Direction des Inventaire et Amenagement (SODEFOR), Abidjan.

Sharma, M. K.: 1986, *Ecofloristic Zones of Africa*, Institute de la Carte Internationale de la Vegetation, Universite Paul Sabatier, Toulouse, France, Distributed by FAO Forest Resources Assessment 1990 Project, Rome, Italy.

United Nations Development Program: 1972, *Forestry Inventory of the Gola Forest Reserves, Report to the Government of Sierra Leone*, FAO.

GREENHOUSE GAS EMISSIONS FROM VEGETATION FIRES IN SOUTHERN AFRICA

R.J. SCHOLES

Division of Forest Science and Technology, CSIR,PO Box 395, Pretoria 0001 South Africa. (bscholes@forestek.csir.co.za)

Abstract. Methane (CH_4), carbon monoxide (CO), nitrogen oxides (NO_x), volatile organic carbon, and aerosols emitted as a result of the deliberate or accidental burning of natural vegetation constitute a large component of the greenhouse gas emissions of many African countries, but the data needed for calculating these emissions by the IPCC methodology is sparse and subject to estimation errors. An improved procedure for estimating emissions from fires in southern Africa has been developed. The proposed procedure involves reclassifying existing vegetation maps into one of eleven broad, functional vegetation classes. Fuel loads are calculated within each 0.5 × 0.5° cell based on empirical relationships to climate data for each class. The fractional area of each class that burns is estimated by using daily low-resolution satellite fire detection, which is calibrated against a subsample of pre- and post-fire high-resolution satellite images. The emission factors that relate the quantity of gas released to the mass of fuel burned are based on recent field campaigns in Africa and are related to combustion efficiency, which is in turn related to the fuel mix. The emissions are summed over the 1989 fire season for Africa south of the equator. The estimated emissions from vegetation burning in the subcontinent are 0.5 Tg CH_4, 14.9 Tg CO, 1.05 Tg NO_x, and 1.08 Tg of particles smaller than 2.5μm. The 324 Tg CO_2 emitted is expected to be reabsorbed in subsequent years. These estimates are smaller than previous estimates.

1. Introduction

Most of Africa has a strongly seasonal climate, with hot, wet summers and warm, dry winters. This kind of climate permits frequent fires, which in turn promote the growth of a fire-tolerant vegetation known as a savanna. Because many savannas are located on extremely infertile soils, frequent application of fire is essential to sustainable low-input pastoralism. Fire has been used for millennia to promote the growth of palatable forage and prevent the encroachment of trees (Goldammer, 1993), and in fact, most fires in Africa are ignited by people rather than by lightning. In the southern hemisphere, savanna fires are concentrated in the dry season, between April and October.

Savanna grass is approximately 45% carbon (C) on a dry-mass basis. When it burns, most of this C is released as CO_2. In general, this does not represent a net release of CO_2, however, because the vegetation regrows in subsequent years. On the other hand, it is a net source of a range of other radiatively active and O_3-forming trace gases and aerosols (particles), because these are not reabsorbed. If the fire frequency or fuel load are increased systematically, vegetation fires are a net source of CO_2, as well; conversely, if fires are suppressed, savannas become a major C sink as tree biomass and organic soil matter increase. The degree to which colonial and post-colonial fire regulations have altered the long-term frequency of fire is unknown. In some places fire may now be more frequent than it was previously, whereas in others, the reverse may be true. African savanna fires

Environmental Monitoring and Assessment **38**: 169-179, 1995.

have a profound effect on the regional and global atmosphere (Crutzen and Andreae, 1990), but there is little data to suggest that the magnitude of the impact has changed in recent times.

The emissions of partial combustion products (CO, CH_4, and aerosols) are inversely proportional to the degree of oxygenation during combustion (Ward and Radke, 1993). The fine, dry grass fuels that predominate in savannas are highly oxygenated during combustion and therefore release relatively small amounts of these gases. Emissions of nitrogenous gases (predominantly NO_x, but also small amounts of nitrous oxide (N_2O) and several other gases) depend on the N content of the fuel (Lobert *et al.,* 1991), which is typically low (0.5–1%). Because of the vast area of Africa that burns, the gross trace gas emissions are globally significant (Crutzen and Andreae, 1990; Levine, 1990; Lacaux, Cachier, and Delmas, 1993) and could dominate the greenhouse gas inventories of large but lightly industrialized African countries, such as Angola, Tanzania, and Zambia.

The general procedure for estimating emissions from biomass burning (which should more properly be called vegetation burning, because biomass refers only to the living material, and fires mostly consume dead material) is to divide the study region into vegetation types and calculate the emission per type (OECD, 1991; UNEP *et al.*, 1995). In each type, the area ($A{:}m^2$) mean fuel load ($L{:}g\ m^{-2}$), combustion completeness (the fraction of the biomass exposed to the fire that actually burns—C:nondimensional fraction), and average time between fires (R:years) is estimated, usually based on expert opinion, because few detailed measurements exist. The amount of biomass consumed per annum ($B{:}\ g\ yr^{-1}$) in each vegetation class is then calculated from $B = A\ R^{-1}\ L\ C$.

The emission factors (F:g emission product/g fuel) are typically estimated by measuring the increase in a given gas relative to the increase in CO_2 in a smoke plume. This measurement is then converted to a dry-mass basis by assuming a C content in the fuel (41–45%) and an efficiency of conversion of fuel C to CO_2 (80–98%). The emission of combustion products is then given for each gas by $E_{gas} = B\ F_{gas}$, which is summed over the various vegetation classes. The method proposed by the Intergovernmental Panel on Climate Change for calculating emissions from savanna burning follows this simple approach (UNEP *et al.*, 1995). The disadvantage of the method is that the range of possible values for R, L, C, and F is very large, and the procedures for validating them in the field are demanding. This means that the accuracy of the estimate is generally very low. This paper briefly describes a method (documented in more detail in Scholes, Kendall, and Justice (1995) and Scholes, Ward, and Justice (1995) that greatly reduces the uncertainties by applying the following strategies:

1. The area burned is measured using a combination of low- and high-resolution satellite images.
2. The fuel load is modeled in four categories, constrained by climate.
3. The emission factors are related to the fuel mixture.

The method has been applied to Africa south of the equator. It could probably also be applied to West Africa and other tropical areas that are subject to frequent and extensive fires, but the predictive equations would have to be recalibrated for areas outside of southern Africa.

The objectives of this paper are to describe a procedure for estimating regional trace gas emissions from vegetation fires and to provide data that will assist in the preparation of national greenhouse gas emission inventories in southern Africa.

2. Methods

The best available vegetation maps (Table I) of southern Africa at an appropriate scale (1:1 M to 1:5 M) were digitized and edge-matched using geographic information system (GIS) technology to form a detailed, composite vegetation map of the subcontinent. The mapped vegetation classes were reclassified into eleven broad types (Table II), each of which has distinctive fuel types and relationships between fuel production and climate. The maps were converted to raster form at a scale matching the available long-term monthly

TABLE I

The best available broad-scale vegetation maps for Africa south of the equator. Where no suitable map is available, the all-Africa map by White (1983) can be used

Country	Source	Scale
Africa	White, F: 1984, Vegetation of Africa, Unesco/AETFAT, Paris.	1: 5 000 000
South Africa, Lesotho, and Swaziland	Acocks, J.P.H.: 1955, Veld Types of South Africa. Memoirs of the Botanical Survey of South Africa 28 (revised 1975 as Memoir 40).	1:1 500 000
Botswana	Ministry of Agriculture, Botswana: 1991, Vegetation map of the Republic of Botswana, Land Utilisation Division, Private Bag 003, Gaborone.	1:2 000 000
Namibia	Giess, W.: 1971, Vegetation Map of South West Africa. Dinteria: 4: 5–114	1:3 000 000
Zimbabwe and Mozambique	Granvaux-Barbosa, L.A.: 1967, Vegetation map of the flora Zambeziaca area, Collins, Harare.	1:2 500 000
Angola	Granvaux-Barbosa, L.A.: 1970, Fitografica de Angola, Instituto de Investicao Cientifica de Angola.	1:2 500 000
Zambia	Edwards, A.C.R.: 1976, Republic of Zambia Vegetation Map, Institut fur Angewandte Geodesia, Frankfort.	1:1 500 000
Malawi	1980, National Atlas of Malawi, Department of Survey, Blantyre.	1:1 000 000
Tanzania	Ministry of Lands: 1984, Tanzania Vegetation Cover Types, Forestry Division, Ministry of Lands, Government of Tanzania.	1:2 000 000
Zaire	1932, Atlas of Southern Katanga, Vegetation Map.	1:500 000
	NASA-GIMMS: In press, Pathfinder Project on Central African Forests.	1:250 000
Kenya	Ministry of Wildlife: 1986, Biotic Communities of Kenya.	1:1 000 000
Uganda	1964 (reissued 1972), Vegetation of Uganda, Department of Land Surveys, Entebbe.	1:500 000

TABLE II

The approximate area of the main vegetation functional types in the countries wholly within southern hemisphere Africa ($m^2 \times 10^9$)

Country	Evergreen forest	Dry forest	Forest/ savanna mosaic	Infertile savanna	Fertile savanna desert	Shrubland/ grassy	Desert	Fynbos	Unpalatable grassland	Palatable grassland	Wetlands
Tanzania	6.1	3.0	0.0	1430.9	132.2	0.0	0.0	0.0	0.0	250.0	33.9
Mozambique	0.0	11.7	11.7	489.1	262.1	0.0	0.0	0.0	0.0	0.0	0.0
Zimbabwe	0.0	0.0	0.0	268.6	121.5	0.0	0.0	0.0	0.0	0.0	0.0
South Africa	2.6	2.6	0.0	50.3	341.9	429.1	0.0	66.4	199.6	130.5	0.0
Botswana	0.0	0.0	0.0	201.3	353.6	0.0	0.0	0.0	0.0	23.1	5.8
Namibia	0.0	0.0	0.0	154.3	500.7	57.3	107.7	0.0	0.0	0.0	0.0
Angola	27.6	8.9	143.9	884.2	155.0	0.0	0.0	0.0	54.1	0.0	0.0
Zambia	0.0	29.9	0.0	605.6	32.9	0.0	0.0	0.0	65.7	0.0	18.1
Malawi	0.0	0.0	0.0	99.1	11.9	0.0	0.0	0.0	0.0	0.0	3.0
Burundi	0.0	0.0	6.2	0.0	12.3	0.0	0.0	37.0	0.0	0.0	0.0
Rwanda	0.0	0.0	0.0	6.2	49.4	0.0	0.0	3.1	0.0	0.0	0.0
Lesotho	0.0	0.0	0.0	0.0	0.0	0.0	0.0	0.0	24.0	5.4	0.0
Swaziland	0	0.0	0	4.3	8.7	0	0	0.0	4.3	0.0	0.0

rainfall surfaces for Africa (Booth *et al.,* 1989; currently a 0.5 × 0.5° grid, soon to be 0.05 × 0.05°). A numerical model calculates the production, herbivory, and decay of the four categories of fuel (green grass, dry grass, tree leaf litter, and fallen twigs) in each month of the year for each raster cell, based on empirical relationships between each of these processes and mean annual rainfall (Scholes, Kendall, and Justice, 1995). There is a linear relationship between rainfall and aboveground grass production, tree litterfall, and tree wood production in all of the vegetation types modeled, but the slope and intercept of the relationship is unique for each. The ratio of monthly rainfall to monthly potential evaporation is used to mimic the greening-up and drying out of the leaves. The fuels are allowed to accumulate over a period equal to the mean period between fires for that cell. The herbivory corrections were calculated based on the relationships between rainfall, soil type, and large mammal carrying capacity published by East (1984). The fuel decay rate depended on moisture and fuel type.

Data from the afternoon overpass of the National Oceanic and Atmospheric Administration (NOAA) satellite was acquired daily for all of southern Africa. The thermal channel (5) of the Advanced Very High Resolution Radiometer (AVHRR) sensor was scanned for pixels that registered significantly higher temperatures than their surroundings, indicating possible fires. Hot pixels that appeared repeatedly (for instance, thermal power plants) were rejected. The number of hot pixels were accumulated per raster cell per month and corrected for the number of days each month that cloud cover obscured the ground. Because the sensor can detect a fire as small as 10 m^2 within a 1 km^2 pixel, the fraction of pixels registering a fire does not directly represent the area burned. Equally, fires that are not active at the time the satellite passes can be missed. The hot pixel count, therefore, requires calibration against a measure of actual area burned.

The actual area burned was visually mapped from fire scars on twelve Landsat MSS images. The images were paired—one from early in the fire season and one from late—and were chosen to form a systematic grid over southern Africa. To test the assumption that fire scars from early in the burning season would still be visible late in the season, the fire scars were traced month by month in three scenes from humid areas (where the scars would have been expected to disappear fastest). The assumption was found to be true. The calibration factor relating the actual area burned to the AVHRR fire count over the same period was found to bear a strongly nonlinear relationship to either the mean annual precipitation or the cumulative Normalised Difference Vegetation Index (NDVI). In arid areas, there are few fires, but they are individually large, because much of the area burns between successive satellite passes, the fraction of the AVHRR pixels that registers a fire is substantially less than the fraction of area burned in one day. In wetter areas, there are many fires smaller than 1 km^2, each of which registers as an entire pixel burned, thus causing an overestimation of the true area.

Emission factors were assembled from the available literature and from field campaigns such as DECAFE and SAFARI '92 (Andreae *et al.,* 1994), held in the savannas of West and southern Africa, respectively. A special campaign was mounted to collect data representative of south Central Africa, where much of the burning occurs (Ward *et al.,* in press). A Fire-Activated Sampling System (FASS) was used to measure the gas concentrations just above the flame zone over the entire duration of a passing flame front. The emission factors were expressed as linear functions of the "combustion efficiency" (the ratio of C released as CO_2 to carbon released as CO_2 plus CO, Ward and Radke, 1993). Combustion efficiency was related to the fraction of the fuel that consisted of dry grass (Ward *et al.*, in press). The mean emission factors in each vegetation type, integrated over month of the year and fuel type, are given in Table III to allow calculation of national emission inventories without recourse to the full model.

3. Results

Table III shows that the broad-leaved, moist savannas on infertile soils are the most extensive vegetation type in southern Africa, followed by the thorny, fine-leaved savannas on fertile soils. Because of their vast expanse, a relatively large fraction of which burns each year (Table IV), these two vegetation types dominate the emissions. Note, however, that the fractional areas burned as predicted by the calibrated satellite images are substantially less than the 33–80% often quoted for savannas (for example, Hao *et al.*, 1990). Similarly, the average fuel loads are lower than is generally suggested (Delmas *et al.*, 1991; Menaut *et al.*, 1991). The grass fuel load is higher on infertile soils than on fertile soils, because grazing is more intense on fertile soils and because fertile soils tend

TABLE III

The mean emission factors for African vegetation types as predicted by the modeling procedure

Vegetation type	CO_2	CO	CH_4	NO_x	Particles (<2.5μm)
	g emitted/kg DM fuel combusted				
Moist/infertile broad-leaved savanna	1523	92	3	6	6
Arid fertile fine-leaved savanna	1524	73	2	5	6
Fynbos (sclero-phyllous thicket)	1519	127	5	9	8
Moist, infertile grassland	1498	59	2	4	5
Arid, fertile grassland	1540	97	3	7	7
Wetlands	1554	58	2	4	5
IPCC (1994) range[a]	1403–1503	67–120	4–7	0.5–0.8	

[a]Assuming 41–45% C content, 85–100% combustion completeness, and the range of emission factors given. Note: The values in this table are weighted by the amount of biomass consumed in the various fuel type categories and therefore integrate over the months of the year and the range of climates covered by each vegetation type. The extensive ecosystems that have been left off this table (principally deserts and semidesert shrublands) burn so infrequently and with such a low fuel load that their emissions can be ignored. Emissions of N_2O are about 10 × less than NO_x and are subject to measurement artifacts.

TABLE IV

The mean mid-dry season fuel loads for various vegetation functional classes in southern Africa, as calculated by an empirical modeling approach driven by long-term rainfall data

Vegetation type	Fuel load on 1 July by type			Fraction of area burned per year
	Tree leaf litter	Twigs	Grass	
	g m^{-2}			
Infertile savanna	23	0	88	0.250
Fertile savanna	44	54	60	0.089
Arid shrubland	0	0	4	0.011
Desert grassland	0	0	8	0.001
Fynbos	109	167	235	0.050
Infertile grassland	0	0	137	0.317
Fertile grassland	0	0	147	0.180
Wetland	0	0	252	0.528

Note: Equations and procedure are described in Scholes, Kendall, and Justice (1995).

to be in drier climates. The presence of trees in the savannas greatly reduces the grass fuel load relative to treeless grasslands, but the reduction in fuel is partially compensated for by tree leaf and twig litter. The grass in fynbos is actually mostly forbs and restios.

The predicted gross pyrogenic emissions for the countries that are wholly within Africa's southern hemisphere are listed in Table V. The total emissions from Africa south of the equator (excluding the island of Madagascar) are 324 Tg yr^{-1} for CO_2, 14.9 Tg yr^{-1} for CO, 0.5 Tg yr^{-1} for CH_4, 1.05 Tg yr^{-1} for NO_x, and 1.08 Tg yr^{-1} for particles smaller than 2.5 μm. (These values exceed the sums of the values in Table V because there are several countries that are partially in the southern hemisphere of Africa but are not included in the table). The 95% confidence limits surrounding the estimates of burned biomass are estimated as $\pm$ 49%, based on an approximation of the root mean square sum of all the component errors. The errors on individual gas emissions will be slightly higher as a result of the additional uncertainty in estimating the emission factors, but they are unlikely to exceed $\pm$60%. The range of variation between the low and high emission factor estimates in the IPCC Guidelines for National Greenhouse Gas Inventories (UNEP *et al.*, 1995) is twofold. The total uncertainty is therefore higher.

4. Discussion

The total emissions calculated by the method described above are significantly lower than previously suggested for African savannas. The only other study in which the emissions can be calculated for exactly the same area is Hao *et al.*, (1990), in which the pyrogenic

TABLE V

Estimated emissions of trace gases and aerosols (particulate matter smaller than 2.5μm in diameter) from southern African countries

Country	CO_2 Tg yr^{-1}	CH_4 Tg yr^{-1}	CO Tg yr^{-1}	NO_x Tg yr^{-1}	N_2O Gg yr^{-1}
Angola	63.5	0.068	2.229	0.160	68.0
Botswana	13.2	0.013	0.434	0.030	7.8
Burundi	3.4	0.006	0.174	0.012	7.7
Lesotho	0.4	0.000	0.011	0.001	0.4
Malawi	7.1	0.007	0.248	0.017	11.7
Mozambique	25.6	0.033	1.035	0.072	40.3
Namibia	4.4	0.004	0.138	0.010	0.3
Rwanda	1.0	0.004	0.090	0.006	1.2
South Africa	11.4	0.015	0.449	0.031	5.8
Swaziland	0.1	0.000	0.000	0.000	0.0
Tanzania	41.0	0.051	1.623	0.114	49.5
Zambia	53.6	0.084	2.502	0.175	73.7
Zimbabwe	7.2	0.008	0.265	0.019	6.7

Note: The reference year is 1989.

emission of CH_4 in southern Africa is estimated to be 4.1 Tg yr^{-1}, in contrast to the 0.5 Tg yr^{-1} estimated here. Other estimates (Delmas *et al.*, 1991; Menaut *et al.*, 1991) are not directly comparable but are higher on a per-unit-area basis than the estimates presented in this paper.

The main reasons for the emissions differences are: (1) the much smaller estimate of the area burned per year, which is the result of using the calibrated satellite measurements rather than anecdotal evidence or uncalibrated coarse-resolution remote sensing; and (2) the lower area-averaged fuel load estimates obtained when primary production is constrained by climate, competition, and herbivory. In addition, the SAFARI field campaign yielded slightly lower emission factors for the gases that are principally produced during the smoldering phase of combustion (CO, CH_4), and this also contributed to the lower estimates. The emission factors for gases generated during the flaming phase (NO_x), on the other hand, have increased.

In the past, estimates of the frequency with which an area burned have generally been based on anecdotal evidence, which can be misleading. From the ground (particularly from roads), it often seems as if every bit of savanna vegetation burns nearly every year, especially in the moister regions of south Central Africa. In fact, some parts of the landscape probably do burn nearly every year, but other parts burn only rarely. The satellite data show that few landscapes burn more than one-third of their area in an average year.

The fuel load estimates that have previously been applied to all savannas are based on a handful of measurements, often derived from favored sites, such as research stations or nature reserves. These places generally have a high fuel load, because they tend to be at the wetter end of the savanna gradient and are lightly grazed. A large part of the savanna area in Africa is semiarid and incapable of producing large fuel loads, except in exceptionally wet years. The moist savannas generally have a high tree cover, which suppresses grass fuel production. The savannas on relatively fertile soils are heavily grazed, which greatly reduces the fuel available for burning.

The IPCC method combines two separate concepts in the term "combustion efficiency": combustion completeness and combustion efficiency (Ward and Radke, 1993). This causes a problem, because the IPCC method then goes on to base the emission factors on CO_2 emissions alone, when the emission factors are highly sensitive to combustion efficiency. Combustion efficiency is generally high in savannas (around 95%) because almost all the fuel is consumed during the flaming phase, which is highly oxygenated. Combustion completeness, on the other hand, is highly variable (50%–98%) and strongly dependent on fuel load, fuel type, and meteorological conditions at the time of the fire. If the fuel load drops below about 50 g m^{-2} of dry grass, the fire becomes very patchy and may go out, especially when the temperature falls and the relative humidity increases at night. Fallen tree leaf litter seldom burns more than 50%, unless there is plenty of dry grass fuel to help it.

It is inappropriate to apply emission factors derived from forest or temperate grassland fires to savanna fires. In general, the emissions of smoldering-phase gases are much lower in savannas, and flaming-phase emissions are higher. This is due to the fact that most of the fuel consists of fine, dry grass, which burns rapidly and efficiently. High emissions in savannas are only associated with large-diameter fuels such as boles or dung.

A key aspect of these results is that the southern African emissions are concentrated in a band that stretches across the subcontinent between about 5° S and 20° S. This belt corresponds to the subhumid, broadleafed savannas on infertile soils. Although the rainfall is sufficient to generate substantial fuel loads, it remains strongly seasonal, permitting frequent fires. The infertility of the soils means that the dry grass from the previous season must be burned to stimulate a flush of palatable new grass.

The method described for estimating emissions from fires in seminatural and natural vegetation will slightly overestimate, because it is based on potential vegetation rather than actual vegetation. This is not an inherent flaw in the method but a reflection of the fact that there is no complete map of actual land cover for Africa. For intensively cultivated countries such as Malawi or Zimbabwe, the estimates will be too high by as much as one-third. For most of southern Africa, the error from this source will be less than 10%.

5. Conclusions

A method has been demonstrated for calculating national- or regional-scale gas (CO_2, CO, CH_4, NO_x) and particle emissions from vegetation burning. Although it is based on the

IPCC method of multiplying the fuel consumed by an emission factor, this method is applied at a fine spatial scale and then summed to give a regional (or national) estimate. The area of vegetation burned is determined from a calibrated satellite remote-sensing method and the fuel load is modeled using climate and vegetation type as drivers. The emission factors used are a function of the fuel type. The method substantially reduces the uncertainties associated with emissions from vegetation fires.

The method produces estimates that are markedly lower (up to an order of magnitude) than previous pyrogenic emission estimates from southern Africa. The reasons for the reduction, in order of importance, are the reduced area-burned estimate, which results from the use of calibrated satellite images in the place of anecdotal evidence, the lower fuel loads when fuel accumulation is constrained by climate and herbivory, and the lower emission factors resulting from recent field campaigns in southern Africa.

Acknowledgments

This research was funded by the U.S. Environmental Protection Agency, the U.S. National Aeronautics and Space Administration (Grant DW80935988–01–0), and the South African Department of Environment Affairs.

References

Andreae, M.O., Fishman, J., Garstang, M., Goldammer, J.G., Justice, C.O., Lovino, J.S., Scholes, R.J., Stocks, B.J., Thompson, A.M., van Wilgen, B., and STARE/TRACE-A SAFARI Science Team: 1994, Biomass burning in the global environment: first results from the IGAC/BIBEX field campaign STARE/TRACE-A/SAFARI-92, in *Global Atmospheric-Biospheric Chemistry*, Prinn, R.G. (ed.), New York: Plenum Press, 83–100.

Booth, T.H., Stein, J.A., Nix, H.A., and Hutchinson, M.F.: 1989, Mapping regions climatically suitable for particular species: an example using Africa, *Forest Ecology and Management* **28**, 18–31.

Crutzen, P.J. and Andreae, M.O.: 1990, Biomass burning in the tropics: impact on atmospheric chemistry and biogeochemical cycles, *Science* **250**, 1669–1678.

Delmas, R.A., Loudjani, P., Podaire, A., and Menaut, J.C.: 1991, Biomass burning in Africa: an assessment of annually burned biomass, in *Global Biomass Burning: Atmospheric, Climatic and Biospheric Implications*, Levine, J.S. (ed.), Cambridge, Massachusetts: MIT Press, 126–132.

East, R.: 1984, Rainfall, soil nutrient status and biomass of large African savanna mammals, *African Journal of Ecology* **22**, 245–70.

Goldammer, J.G.: 1993, Historical biogeography of fire: tropical and subtropical, in *Fire in the Environment: The Ecological, Atmospheric and Climatic Importance of Vegetation Fires*, Crutzen, P.J., and Goldammer, J.G. (eds.), Dahlem Workshop Reports, Environmental Sciences Research Report 13, Chichester: John Wiley & Sons, 298–314.

Hao, W.M., Liu M.H., and Crutzen, P.J.: 1990, Estimates of annual and regional releases of CO_2 and other trace gases to the atmosphere from fires in the tropics, based on the FAO statistics for the period 1975–1980, in *Fire in the Tropical Biota: Ecosystem Processes and Global Challenges*, Goldammer, J.G. (ed.), Ecological Studies 84, Berlin: Springer-Verlag, 440–462.

Lacaux, J.P., Cachier, H., and Delmas, R.: 1993, Biomass burning in Africa: an overview of its impact on atmospheric chemistry, in *Fire in the Environment: The Ecological, Atmospheric and Climatic Importance of Vegetation Fires*, Crutzen, P.J., and Goldammer, J.G. (eds.), Dahlem Workshop Reports, Environmental Sciences Research Report 13, Chichester: John Wiley & Sons, 159–192.

Levine, J.S.: 1990, Global biomass burning: atmospheric, climatic and biospheric implications. *EOS* **71**, 1075–1077.

Lobert, J.M., Scharffe, D.H., Hao, W.-M., Kuhlbusch, T.A., Seuwen, R., Warneck, P., and Crutzen, P.J.: 1991, Experimental evaluation of biomass burning emissions: nitrogen carbon containing compounds, in *Global Biomass Burning: Atmospheric, Climatic and Biospheric Implications*, Levine, J.S. (ed.), Cambridge, Massachusetts: MIT Press, 289–304.

Menaut, J.C., Abaddie, L., Lavenu, F., Loudjani, P., and Podair, A.: 1991, Biomass burning in West African savannas, in *Global Biomass Burning: Atmospheric, Climatic and Biospheric Implications*, Levine, J.S. (ed.), Cambridge, Massachusetts: MIT Press, 133–142.

Organisation for Economic Co-operation and Development (OECD): 1991, Estimation of greenhouse gas emissions and sinks. Paris, France: OECD.

Scholes, R.J., Kendall, J., and Justice, C.O.: 1995, Biomass burning in southern-hemisphere Africa, *Journal of Geophysical Research*. in press.

Scholes, R.J., Ward, D., and Justice, C.O.: 1995, Trace gas emissions from biomass burning in southern-hemisphere Africa, *Journal of Geophysical Research*. in press.

UNEP, OECD, IEA, IPCC (United Nations Environment Programme, Organisation for Economic Co-operation and Development, International Energy Agency, Intergovernmental Panel on Climate Change): 1995, IPCC *Guidelines for National Greenhouse Gas Inventories*, IPCC, Bracknell, 3 Volumes.

Ward, D.E., Hao, M., Susott, R.A., Babitt, R.A., Shea, R.W., Kauffman, J.B., and Justice, C.O.: 1995, Effect of fuel composition on combustion efficiency and emission factors for African savanna ecosystems, *Journal of Geophysical Research*. in press.

Ward, D.E. and Radke, L.F.: 1993, Emission measurements from vegetation fires: comparative evaluation of methods and results, in *Fire in the Environment: The Ecological, Atmospheric and Climatic Importance of Vegetation Fires*, Crutzen, P.J., and Goldammer, J.G. (eds.), Dahlem Workshop Reports, Environmental Sciences Research Report 13, Chichester: John Wiley & Sons, 53–76.

White, F.: 1983, *The Vegetation of Africa*, Natural Resources Research XX, Paris: Unesco, 356.

DETERMINATION OF BIOMASS BURNING EMISSION FACTORS: METHODS AND RESULTS

R. DELMAS, J.P. LACAUX, and D. BROCARD

Laboratoire d'Aérologie (URA CNRS 354), Université Paul Sabatier, 118 Route de Narbonne 31062, Toulouse Cedex, France

Abstract. Biomass burning, in a broad sense, encompasses different burning practices, including open and confined burnings, and different types of vegetation. Emission factors of gaseous or particulate trace compounds are directly dependent both on the fuel type and the combustion process. Emission factors are generally calculated by stoichiometric considerations using the carbon mass balance method, applied either to combustion chamber experiments or to field experiments based on ground-level measurements or aircraft sampling in smoke plumes. There have been a number of experimental studies in the last 10 years to investigate wildfires in tropical, temperate, or boreal regions. This article presents an overview of measurement methods and experimental data on emission factors of reactive or radiatively active trace compounds, including trace gases and particles. It focuses on fires in tropical regions, that is, forest and savanna fires, agricultural burns, charcoal production, use of fuelwood, and charcoal combustion.

1. Introduction

Although biomass burning by humans is one of the most ancient forms of anthropogenic atmospheric pollution, its importance in determining the chemical composition of the atmosphere was pointed out only 15 years ago (Seiler and Crutzen, 1980). During the 1980s much research was conducted, especially in the tropics, to study in detail trace gas and particle emissions from vegetation fires in various ecosystems. The pioneering, but also very preliminary, experiments were conducted in South America (Crutzen *et al.*, 1979) and Africa (Delmas, 1982). They were followed by more detailed investigations in these two continents and also in Australia (Crutzen *et al.*, 1985; Ayers and Gillett, 1988; Bonsang *et al.*, 1991). At the beginning of the 1990s, important international campaigns were launched: BASE-A and BASE-B in Brazil (forest and savanna fires) and FOS/DECAFE in Ivory Coast (savanna fires). The most recent study, STARE (TRACE-A, SAFARI, 1992), focused on savanna fires in southern Africa to document not only trace compound emissions but also their influence on regional and global atmospheric chemistry. This campaign gathered some 200 scientists in Africa, South America, and over the tropical Atlantic. Several books have been published recently to review the knowledge on biomass burning and its atmospheric and ecological consequences (Goldammer, 1990; Levine, 1991; Crutzen and Goldammer, 1995).

Under ideal conditions, that is, complete combustion, the burning of organic material is an oxidation process that primarily produces water vapor and carbon dioxide (CO_2), according to the simplified reaction:

$$(CH_2O) + O_2 \quad CO_2 + H_2O$$

Environmental Monitoring and Assessment **38**: 181-204, 1995.

where (CH_2O) represents the average composition of organic matter. However, in natural fires, because the oxygen (O_2) supply is never sufficient, there is incomplete combustion. This leads to the formation of incompletely oxidized compounds (carbon monoxide, or CO), or reduced compounds such as methane (CH_4), nonmethane hydrocarbons (NMHC), and ammonia (NH_3), among others.

The emission of any gaseous or particulate emission from biomass burning depends on two parameters: the amount of biomaterial burned and the proportion of a given compound released during the combustion (the emission factor, or EF). The EF is generally calculated by reference to CO_2 on the basis of an experimental emission ratio: $ER=\Delta[X]/\Delta[CO_2]$, corresponding to the excesses of these two constituents in the fire plume compared with the average concentration in ambient air.

The emission ratio (ER) and its variability, is critical for any estimate of trace compound emissions from biomass burning at local, regional, or global levels. The ER is highly dependent on the type of combustion, especially the relative importance of the two main phases of the combustion, flaming and smoldering. Whatever the combustion phase, the main compound emitted is CO_2, which represents between 85% and 90% of the carbon (C) released into the atmosphere (Lobert *et al.*, 1991). However, as noted in many papers (for example, Crutzen and Andreae, 1990; Andreae, 1991), this emission does not correspond in all cases to a net C loss from the terrestrial biosphere, for example, savanna fires and agricultural waste burnings. The CO_2 released by these combustions is removed from the atmospheric reservoir and reincorporated back into the biomass via photosynthetic activity during the vegetation regrowth within weeks or months after burning. In this case, biomass burning can be seen as an abiotic equivalent of the bacterial decomposition of the organic material. On the other hand, tree burning through deforestation and the unsustainable use of wood as fuel constitute a net loss of C from the terrestrial biosphere and thus contribute to the additional greenhouse effect. However, compared with the natural process of aerobic decomposition of organic matter in soil, almost all open and confined burning represents net additional sources of radiatively active or reactive trace compounds [CH_4, nitrogen oxide (N_2O), ozone (O_3) precursors, and aerosols].

This article provides an overview of the present knowledge on trace gas and aerosol EFs, which are one of the basic parameters required to calculate emissions from biomass burning. The different types of biomasses exposed to fire and the variability of EFs in relation to the burning processes are briefly discussed. Based on the numerous field campaigns and laboratory measurements, a table of estimates of EFs is presented for the most important trace compounds resulting from the dominant types of combustion in tropical regions.

2. Biomass burning types

Broadly, biomass burning encompasses different burning practices and different types of vegetation. A distinction can be made first between natural (accidental) and anthropogenic burns (which are largely dominant), between open and confined fires, and between tropical and temperate-boreal regions (Ahuja, 1994). Confined fires include all types of domestic or industrial burning with any of various combustion devices, while open fires include fires in grasslands, croplands, and forests. In tropical regions the dominant practices are savanna and forest fires, use of wood as fuel, and agricultural burns, whereas in temperate and boreal regions of the Northern Hemisphere, forest fires and agricultural waste burning dominate (Dixon *et al.*, 1994). According to several authors (Levine, 1991; Andreae, 1991), about 80% of global biomass burning occurs in the tropics. A detailed description of burning practices throughout the world can be found in various papers (Seiler and Crutzen, 1980; Crutzen and Andreae, 1990; Andreae, 1991).

Almost all the various burning activities in tropical regions are anthropogenic. The major factors driving these activities are linked to the need of developing countries for resources to support growing populations. Forest clearing for permanent agriculture and ranching, shifting cultivation, and the use of wood as fuel are examples of such activities.

Savanna fires are a very old and widespread phenomenon. The savanna area burned each year is about 7.5 million square kilometers (km^2) (Hao *et al.*, 1990). In Africa, more than 70% of savanna is potentially exposed to fire every year. As a consequence, Africa is the world's leading contributor to pollution originating from the burning of vegetation (Hao *et al.*, 1990; Cachier, 1992).

Forest clearing does not affect tropical rainforests to the same extent on each continent (Brown *et al.*, 1993). Each year, 0.6% of the forested area is expected to be destroyed and burned, with presently half of the deforestation taking place in South America. Significant tropical deforestation began at the end of the last century, but it has dramatically increased during the last 30 years as a result of population increases. For example, in western Africa, 70% of the forested area that existed at the beginning of the century has already been cleared.

Shifting cultivation, a cyclical practice of cultivation and a fallow period on the same plot, is often accompanied by biomass burning. Shifting cultivation is thought to be practiced by 200 million people in the world, mainly in tropical countries, with an annual clearing of 20 to 60 million hectares (Mha) (Seiler and Crutzen, 1980; Hao *et al.*, 1990).

The removal of wood for fuel affects both forest and savanna ecosystems. Fuelwood is the main source of energy for domestic use in most tropical countries. It can be used directly for cooking or heating, or it can be made into charcoal, primarily to supply large cities with a domestic energy source. Charcoal production is increasing as a result of urban growth in developing countries.

The last significant source of tropical biomass burning are seasonal agricultural burns. The primary crops burned are rice straw in Southeast Asia and sugar cane, which is almost entirely burned on fields before harvesting and can be a very important local source of pollution (Kirchhoff *et al.*, 1989).

3. Emission ratio (ER) and emission factor (EF)

The calculation of gaseous or particulate trace compound emissions from biomass burning is based on the concepts of emission ratio (ER) and emission factor (EF). These concepts are necessary because absolute concentrations in smoke plumes represent only the degree of dilution of combustion products in ambient air. The emissions from biomass burning are proportional to the amount of dry matter actually burned (M), and to the emission factor (EF[X]) of each compound (X), which represents the quantity emitted per unit of dry matter (or C) burned. This factor is generally expressed in g(X)/kg of dry matter (or C) burned. Knowledge of M (dry matter burned per unit area for open fires, or per unit time for confined burning) allows the emission flux to be calculated. Unfortunately, the EF[X] cannot be calculated from direct concentration measurement of X in smoke plumes. The EF is derived from the ER using the carbon mass balance method, because the amount of C in the biomass material (around 45%) does not vary significantly among biomass types, and most of the carbonaceous species can be measured. In this method, ERs are calculated by reference to CO_2 or carbon monoxide (CO) using the regression line between the species of interest and these two gases, which are the dominant C compounds emitted whatever the type of combustion (Table I). The total amount of C involved in the combustion process includes the C left as combustion residue (charcoal and ashes) and released by combustion as CO_2, CO, CH_4, NMHC, oxygenated organics, and particulate C.

The selection of CO_2 or CO depends on the type of combustion, which is linked to the characteristics of the vegetation for open fires and to the characteristics of the combustion devices for confined fires. The type of combustion, or combustion device, determines the relative importance of the two main phases of combustion, flaming and smoldering. It has been shown, for instance, that the EF of CH_4 may vary by more than a factor of 10 among savanna fires, forest fires, and charcoal production (Delmas *et al.*, 1991a). Clearly, reduced or incompletely oxidized compounds are preferentially emitted during the smoldering phase. For these compounds, the correlation is generally better with CO than with CO_2. Figure 1 illustrates this phenomenon for the emission of CO_2, CO, and CH_4 from grass burning in a combustion chamber. Therefore, for detailed case studies on emissions from biomass burning, the choice of the reference gas is important to accurately calculate emission fluxes.

Based on the C mass balance method, the classical definition of the EF, described in detail in Hurst *et al.*, (1994a and b), is as follows:

$$EF[CO_2] = \frac{\dfrac{\Sigma C_{emit}}{C_{fuel}}}{1+\dfrac{\Delta CO}{\Delta CO_2}+\dfrac{\Delta CH_4}{\Delta CO_2}+\dfrac{\Delta \Sigma NMHC}{\Delta CO_2}+\dfrac{\Delta PC}{\Delta CO_2}}$$

TABLE I

Molar emission ratios $\Delta X/\Delta CO_2$ and $\Delta X/\Delta CO$ for the main trace gas species emitted by savanna fires, as measured during field campaigns and laboratory experiments

* $ER/CO_2 = \Delta X/\Delta CO_2$ * $ER/CO = \Delta X/\Delta CO$	Australian fires[a]		FOS/DECAFE[b]	SAFARI[c]	Combustion chambers[d]	
	ER/CO_2 10^{-3}	ER/CO %	ER/CO_2 10^{-3}([e])	ER/CO_2 10^{-3}([e])	ER/CO_2 10^{-3}	ER/CO %
CO_2	1 000		1 000	1 000	1 000	1 380
CO	90±26	100	61±15	62±10		100
CH_4	4.2±1.9	4.5±0.9	3.1±0.5	4±0.5		9.11
Σ NMHC	9.9±4.1	19±10	7.0±1.5	6±2		25.41
N_2O	0.02±0.02	0.1±0.1	0.14±0.05	0.09	0.09	
NO_X	2.1±0.4	2.3±1.0	1.37±0.10	2.8±0.3	2.0	
NO_Y			1.87±0.15	3.5±0.4		
NH_3	2.9±1.9	2.6±1.0	0.057±0.5	0.7±0.4		1.55
N_2					4.7	
HCN	0.023±0.02	0.028±0.03				1.13
CH_3CN	0.068±0.05	0.079±0.03				0.25
H_2				10		
SO_2			0.047±0.02	0.25		
COS			0.0017±0.0015	0.01		
Σ S	0.016	0.35±0.36				
CH_3Cl[f]				0.95		0.16
CH_3Br[f]				0.0083		
CH_3I[f]				0.0026		

[a]from Hurst *et al.*, 1994b
[b]from Lacaux *et al.*, 1993
[c]from Andreae *et al.*, 1995
[d]from Lobert *et al.*, 1991
[e]best-guess estimates from all FOS/DECAFE and SAFARI data
[f]relative to CO

where *SCemit* is the mass of C released into the atmosphere during a fire and *Cfuel* is the mass of fuel C burned. *SCemit/Cfuel* represents the fraction of burned fuel C released into the atmosphere during the combustion, and $\ddot{A}X/\ddot{A}CO_2$ represents average molar emission ratios of other C species relative to CO_2:

$$ER[X] = \frac{\Delta X}{\Delta CO_2} \qquad (1)$$

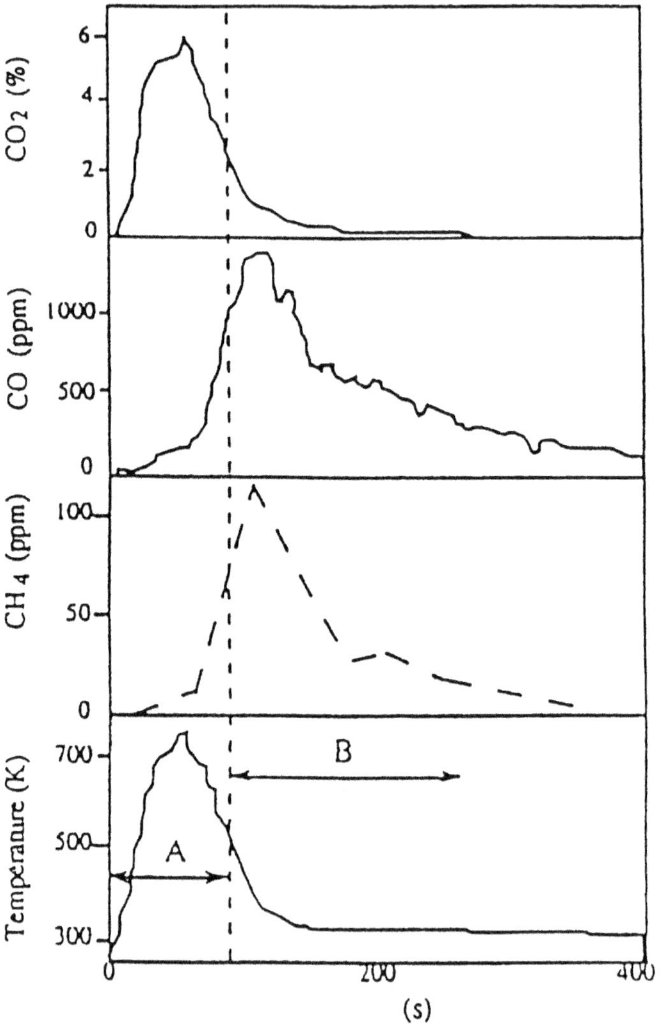

Figure 1. Concentrations of CO_2, CO, and CH_4 in smoke, and temperature of an experimental fire (grasses in a combustion chamber) as a function of time. The dashed line separates the two phases of combustion, flaming and smoldering (from Crutzen and Andreae, 1990).

where

$D[X] = [X]_{smoke} - [X]_{ambient}$, and

$D[CO_2] = [CO_2]_{smoke} - [CO_2]_{ambient}$ (3)

Thus, EFs for those species are given by the following equations for flaming stage and smoldering stage (3) emissions:

$$EF[X] = \frac{\Delta X}{\Delta CO_2} n EF[CO_2]$$

$$EF[X] = \frac{\Delta X}{\Delta CO} n EF[CO]$$

where n is the number of C atoms per molecule of the species X.

Using these equations, the sum of EFs for C-based trace gas and particulate C obey the relationship $\Sigma EF[X] = \Sigma Cemit/Cfuel$, and the sum of EFs of all carbonaceous species, plus the C remaining in ashes, must be unity.

The same equation can be used to calculate EFs for nitrogen- (N) or sulfur-based species. Equations 2 and 3 should be divided by the mean molar N/C (1.16) or S/C (2.66) ratio of the fuel. The mean molar fuel N/C ratio can be calculated by dividing the mean N load by the mean C load and dividing by the respective atomic weights. The EF for an N- (sulfur) based species X calculated this way therefore represents the fraction of burned fuel N (S) emitted as species X, and is independent of the fuel N (S) content.

Ward *et al.* (1992), during the BASE-B experiment, used another approach based on linear models relating EFs to combustion efficiency. The combustion efficiency representing the degree of oxidation of the fuel (proportion of C oxidized as CO_2) should be calculated from the complete C mass balance. However, a good approximation is obtained from the ratio $CO_2/[CO_2+CO]$. The sum $[CO_2+CO]$ represents generally more than 95% of the C released by combustion. Ward *et al.* calculate the combustion efficiency by multiplying the C concentration above background, as measured using CO_2 and CO real-time sensors ($g[C]/m^3$), by the vertical wind velocity (m/s), and this by the ratio of fuel to C. Two regression models were obtained for methane for savanna and forest fires:

Savanna (Cerrado): $EF[CH_4] = 32.05 - 32.53(CE)$ $R^2 = 0.87$

Deforestation: $EF[CH_4] = 63.41 - 64.69(CE)$ $R^2 = 0.91$

Such models should be tested in other tropical ecosystems.

For global estimates, in which only the amount of biomass burned is known, the simplest approach, based on an equation such as the following, is suggested:

where

$\Sigma[X]$ represents the emission flux of the compound X.

B represents the biomass exposed to fire.

α represents the fraction of the biomass that is actually burned, also called combustion completeness.

$EF[X]$ represents the EF of the compound X ($g\ kg^{-1}$ dry fuel).

Emissions from open or confined fires can be calculated the same way with appropriate estimates of biomasses actually burned (αB). Average EFs should be considered from best-guess estimates for each fire type (see "Experimental data on emission ratios and emission factors for tropical biomass burning"). If the nitrogen (N) content of the fuel is known, EFs should be calculated from average emission and fuel C/N ratios.

Based on the above considerations, the information needed to estimate trace gas and particulate emissions from biomass burning relies on (1) chemical measurements of ERs in smoke plumes and fuel composition (C, N, sulfur (S) content), allowing the EFs to be calculated, and (2) estimates of biomasses submitted to fire and biomasses actually burned. The following sections present an overview of experimental data on measurements of ERs and EFs for the main burning processes in tropical regions.

4. Experimental methods for molar-emission ratio determination

The experimental determination of an ER for a given chemical species X is based on simultaneous concentration measurements in smoke plumes of X and of the reference gas (CO_2 and/or CO). Measurements should be taken with the same integration time: from a few seconds or less for most trace gases up to a few minutes for atmospheric aerosols, or from the same sample when samples are taken in canisters or teflon bags. This last method is used for laboratory measurements by gas chromatography or by any other analytical technique that cannot be deployed in the field.

A number of analytical techniques have been used in recent field experiments. They are described in the corresponding papers; details can be found in Ward and Radke (1993). Generally, CO_2 and CO are measured by nondispersive infrared analysis (NDIR). Gas chromatography with flame ionization detection is the most common technique for CH_4 and NMHC. Total NMHC is measured in the field, while speciated NMHC or oxygenates are measured in the laboratory from grab samples taken in smoke plumes. N compounds (NO_x, NO_y, NH_3) are usually measured by chemiluminescence; N_2O and S or halogenated compounds are measured by gas chromatography with different types of detectors (electron capture detector, photometric detector).

A number of other techniques have also been used for trace gases: liquid sampling and ionic chromatography for HNO_3 or organic acids measurements (Lacaux *et al.*, 1993), gas phase Fourier transform infrared spectroscopy (Griffith *et al.*, 1991; Hurst *et al.*, 1994a), and tunable diode laser (Harris *et al.*, 1993; 1995). The particle number/size distribution is measured by laser-optical particle counters. The black C content can be monitored by continuous optical measurements (aethalometer) (Cachier *et al.*, 1995), while the chemical composition (organic and mineral component) of aerosol is determined by different techniques, including PIXE (proton-induced x-ray emission) (Maenhaut *et al.*, 1995), electron microscopy (Gaudichet *et al.*, 1995), capillary electrophoresis (Andreae *et al.*, 1995a,b,c), thermochemical analysis (Cachier *et al.*, 1995), and gas or liquid chromatography/mass spectrometry (Clain, 1995).

For open fires, sampling techniques can be divided into two distinct families: those associated with real-time concentration measurements in the smoke and those associated with laboratory measurements. For real-time concentrations, the analytical instruments must be close to the fire. Several platforms are used; a distinction can be made between ground-level and airborne measurements.

Several systems have been used to take field measurements of ground-level trace gas and aerosol concentrations in smoke plumes. The fire atmospheric sampling system (FASS) (Ward *et al.*, 1992) consists of a mast set up on experimental plots several hours before a fire. The mast is equipped with temperature and wind probes and gas and particulate matter (less than 2.5 μm diameter, or PM 2.5) samplers connected to real-time gas (CO_2, CO, NO) sensors or canisters placed in a computer-driven data logger module that is inserted into the soil at the bottom of the mast. FASS, which has been used in South America (BASE-B) and Africa (SAFARI), provides a single measurement of the whole fire process on each studied plot, and can be used for any type of fire, even very intense ones. Another system, used in the FOS/DECAFE and SAFARI experiments (Lacaux *et al.*, 1993; 1995), gives more continuous measurements because it follows the fire front. It is based on handheld sampling devices supplied by tubing held on long poles and connected to real-time gas analyzers, or by aerosol samplers supported by a mobile platform. Canisters or teflon bags for laboratory analyses can be filled using the same system.

Aircraft sampling, either in smoke plumes close to the fires (Helas *et al.*, 1995; Atlas *et al.*, 1995) or in polluted layers (Delmas *et al.*, 1991a; Marenco *et al.*, 1990), has also been extensively used. Various aircraft have been used in the last campaigns in the tropics: a small, one-engine aircraft during FOS/DECAFE; larger, two-engine aircraft (a Bandeirante EMB101 for BASE-B and a Douglas DC–3 for SAFARI); and even a jet (Caravelle) for the TROPOZ II experiment (Marenco *et al.*, 1992). Helicopters were also shown to be very well adapted to this type of measurement (Cofer *et al.*, 1995). For ground-level experiments, real-time gas-phase analytical instruments are set up on board aircraft or helicopters and complemented by aerosol or gas samplers for laboratory analyses.

Airborne and ground-level sampling techniques generally yield quite compatible results, as illustrated in Figure 2a, 2b (from Bonsang *et al.*, 1995), which shows the relationship between excess CH_4 and CO or C_2H_2 and CO. This figure is based on several data sets from ground-level and aircraft sampling in canisters or teflon bags. A similar comparison of sampling techniques was made by Andreae *et al.* (1995a) for SAFARI data. These authors show that $\Delta CO/\Delta CO_2$ or$\Delta NO_x/\Delta CO_2$ ratios measured by the various groups were quite similar no matter what the sampling procedure. In combustion chamber measurements, ratios measured CO/CO_2 are often higher than in open fires with the same type of biomass. The difference is not due to the sampling procedure but rather to a difference in the combustion process itself that favors the smoldering phase, because the chamber does not perfectly reproduce natural open-fire conditions.

5. Experimental data on emission ratios and emission factors for tropical biomass burning sources

As already mentioned, about 80% of biomass burning sources are in the tropics, with a significant percentage in Africa. Several attempts to estimate biomass burning sources in

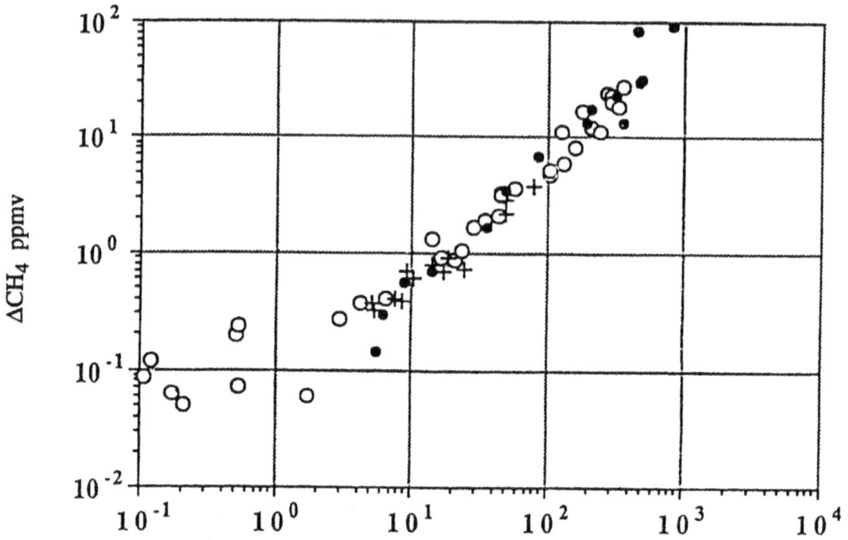

Figure 2a. Excess CH_4 versus CO in savanna fire samples. Open circles represent flaming samples (canisters); closed circles, smoldering samples (canisters); crosses, integrated samples (tedlar bags) (from Bonsang *et al.*, 1995).

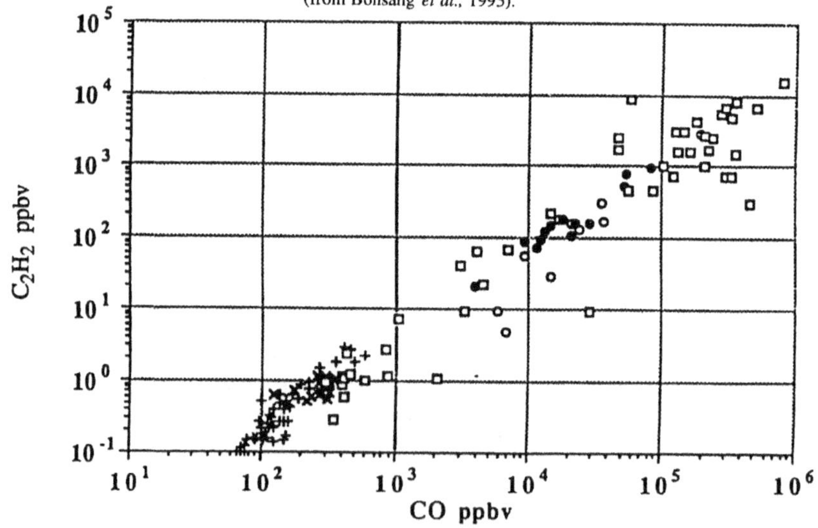

Figure 2b. C_2H_2 versus CO in FOS/DECAFE samples and for the background air over Africa. Open circles represent 1989 FOS experiment (canisters); open squares, 1991 FOS experiment (canisters); closed circles, 1991 FOS experiment (integrated ground-level samples); crosses, aircraft measurements in and above the boundary layer. Results taken from DECAFE 88 experiment, TROPOZ I and TROPOZ II over tropical and equatorial Africa (from Bonsang *et al.*, 1995).

Africa have already been made (Delmas *et al.*, 1991b; Hao *et al.*, 1990; Hao and Ward, 1993). More detailed investigations are needed to improve these preliminary estimates; however, it seems that the order of importance, from greatest to least, of the various types of biomass burning could be as follows: savanna fires, firewood burning (including charcoal burning and production), forest fires, and agricultural waste burning. In spite of significant discrepancies among the various estimates, the different authors agree on the overwhelming importance of savanna fires in Africa. The atmospheric effect of savanna fires has been described in a number of papers dealing with ozone formation over equatorial Africa and the tropical Atlantic (Cros *et al.*, 1988; Fishman *et al.*, 1990, 1991; Andreae *et al.*, 1992).

The review in this article of experimental data on EFs focuses on those species that have a direct or indirect radiative impact (greenhouse gases, ozone precursors, and atmospheric aerosols). Information on the ER of minor species like speciated hydrocarbons or oxygenates, halogenated sulfur and nitrogen compounds, and the particulate organic C composition will complement this data inventory.

5.1. Savanna Fires

Three complete data sets were obtained from prescribed fires in Australia (Hurst *et al.*, 1994a and b), the FOS/DECAFE experiment in the Ivory Coast (Lacaux *et al.*, 1993; 1995), and the SAFARI campaign in South Africa (Cofer *et al.*, 1995; Andreae *et al.*, 1995a,b,c). These field data were compared with combustion chamber data published a few years ago (Lobert *et al.*, 1991).

The most recent data set, from the SAFARI campaign, is the most complete and precise because it concerns measurements carried out by several groups in different ecosystems. Ratios, $\Delta CO/\Delta CO_2$ (mol%) measured in the various ecosystems range from 3.5±1.0 in semiarid savanna in Zimbabwe (Le Canut *et al.*, 1995) to 9.4±1.3 in a mixed forest-shrub area in the humid montane area of Natal, South Africa (Koppmann *et al.*, 1995). As previously discussed, the average $\Delta CO/\Delta CO_2$ ratio is a good indicator of the quality of the combustion; a low value is associated with intense flaming conditions. An average value of 6.2% is considered as representative of average savanna fires (Andreae *et al.*, 1995a). ERs of CH_4 and total NMHC are fairly well in accordance with $\Delta CH_4/\Delta CO_2$= 0.31 – 0.42 % and ΔNMHC/ΔCO_2= 0.6 – 1.0 %, respectively. These values are similar to those previously published: 0.6% (CH_4) and 1.1% (NMHC) for Brazilian savanna fires (Greenberg *et al.*, 1984), and 0.54% (CH_4) and 0.81% (NMHC) for humid savanna in the Ivory Coast (Bonsang *et al.*, 1991). There were no significant differences among C content; the small discrepancies observed between different values are probably due to differences in vegetation structure and plant water content, both of which affect the quality of the combustion (high humidity, which promotes the smoldering phase, leads to enhanced values of CH_4 and NMHC).

ERs are highly dependent on the chemical composition of plants. This is especially true for N compounds, because the N content of plant material may vary according to the type of plant, and also according to the season for a given plant type (Delmas *et al.*, 1995). Figure 3 shows a linear relationship between the NO_x ER and the N content of plant material

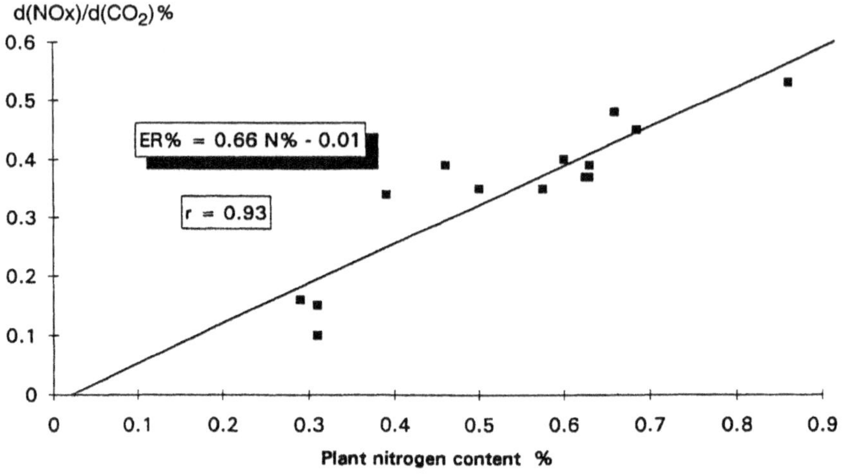

Figure 3. NO_x emissions by savanna fires: linear relationship between molar emission ratio and plant nitrogen content (from Lacaux *et al.*, 1995).

based on two data sets (FOS/DECAFE and SAFARI) obtained by the same technique (Lacaux *et al.*, 1995). All data sets confirm that biomass burning should not be considered a significant source of N_2O. For ammonia, more measurements are needed. Data from Australia and SAFARI agree quite well, with an ER of about 0.2%.

ERs of sulfur compounds are rather low, which is in accordance with the very low sulfur content of the terrestrial vegetation. Conversely, SAFARI data show that biomass burning is a significant source of methyl chloride (CH_3Cl).

Biomass burning aerosols are believed to have a significant impact on the Earth's radiative budget, directly through their own radiative properties, and indirectly through their role as cloud condensation nuclei (CCN). Particulate C is an important product of savanna fires. There is evidence that the major part of this aerosol is composed of polymerized organic species, of which the most refractory material forms the black C fraction (also called soot carbon). The black C content of the particles (3% to 25%) is highly variable and depends primarily on the physical parameters of the burns (Cachier *et al.*, 1991). Emitted particulate C may represent about 1% of the fuel C. An example of fuel C and N partitioning during a savanna fire in Australia (Hurst *et al.*, 1994b) is shown in Figure 4. The proportion of undetected N represents 43% of the total N content of the fuel biomass. A similar imbalance of 50% has been observed in the FOS/DECAFE experiment (Delmas *et al.*, 1995). These results agree with combustion chamber data (Lobert *et al.*, 1991). The emission of molecular N, the so-called pyrodenitrification

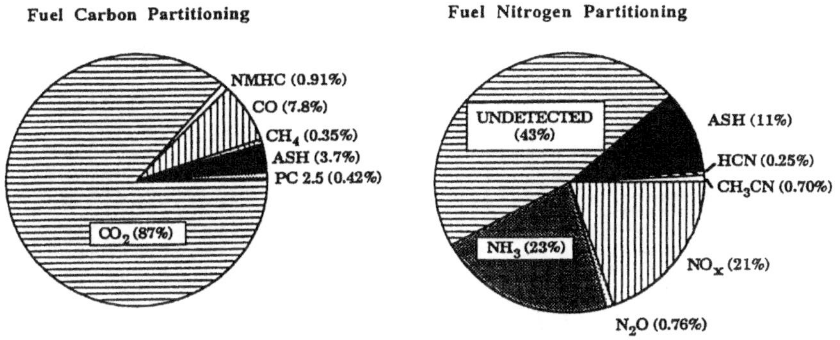

Figure 4. Partitioning of fuel C and N burned during savanna fires at Kapalga (Australia). PC 2.5 is emitted particulate carbon (<2.5 μm diameter), as measured by Ward *et al.* (1992) for Brazilian cerrado fires (from Hurst *et al.*, 1994b).

(Kuhlbush *et al.*, 1991), would represent approximately half of the N emissions from biomass burning.

Carbonaceous matter in gas or aerosols exists in a considerable diversity of products, including alkanes, alkenes, aldehydes, ketones, furans, esters, organic acids, polycyclic aromatic hydrocarbons (PAHs), and other compounds (Koppmann *et al.*, 1995; Atlas *et al.*, 1995). Figure 5a shows an example of the repartition between black C (Cb) and total C (Ct) in rainwater samples taken at Lamto, located in the humid savanna of Ivory Coast. There is a large increase in Cb and Ct concentrations during the dry season (November-March) due to savanna fires, with a range of 20% to 30% for the Cb/Ct ratio (*Cachier et al.*, 1991). Figure 5b shows the relative distribution of organic compounds as a function of the number of C atoms in aerosol samples of savanna fires taken during the SAFARI experiment (Clain, 1995).

Based on all published measurements of savanna fires, and especially on the most recent ones (SAFARI), Andreae *et al.* (1995a) propose a best-guess estimate of the EFs for savanna fires expressed in g(species)/kg dry fuel, which can be used, on any geographical base, to directly calculate trace gas and particulate emissions from savanna fires (Table II).

5.2. Firewood and Charcoal

Biomass fuels are used by about half the world's population as the major source of domestic energy. Indoor air pollution caused by biomass fuel is a potentially serious problem studied by the World Health Organization. The contribution of pollutants from biomass burning to the chemical composition of the atmosphere is undoubtedly significant, even though there are only a very limited number of studies.

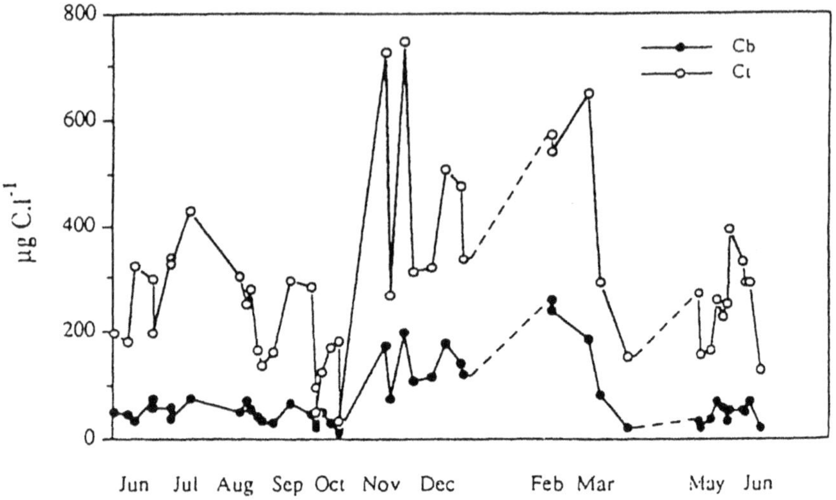

Figure 5a. Seasonal variations of black carbon (Cb) and total carbon (Ct) in rainwater samples taken at Lamto, Ivory Coast.

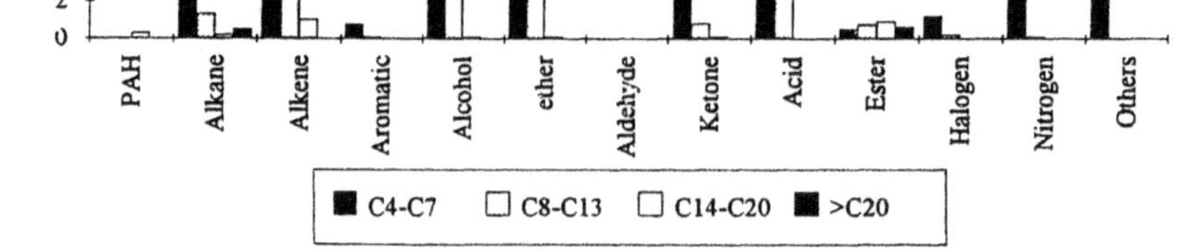

Figure 5b. Relative distribution of organic compounds as a function of carbon atom number in the aerosol produced during the flaming-phase combustion of *Loudetia simplex* (savanna grass) in a combustion chamber (from Clain, 1995).

TABLE II

Estimates of emission factors for trace gas and aerosols for savanna fires

	Gases						
	Greenhouse gases and O_3 precursors						
	CO_2	CO	CH_4	NMHC	NO_X	N_2O	CH_3Cl
EF g kg^{-1} dry fuel	1 640	65	2.4	3.1	3.1	0.15	0.11
	Other gases species						
	NO_Y	NH_3	SO_2	H_2	COS	CH_3Br	CH_3I
EF g kg^{-1} dry fuel	3.9	1	0.6	0.75	0.02	0.002	0.001
	Aerosols						
	Species						
		TPM	PM 2.5	C	BC	K	CCN
EF g kg^{-1} dry fuel		10	5	7	0.8	0.33	$1.1*10^{12}$

Source: Andreae *et al.*, 1995a

The first results of a study on C and N-compound emissions from biomass fuel burning in West Africa have been recently published (Lacaux *et al.*, 1994; Brocard *et al.*, 1995), and are summarized in Table III. This study includes two field campaigns in the Ivory Coast. The first one was conducted in April 1992: Gases (CO, CO_2, CH_4, NMHC, organic acids, NO_x, NH_3, N_2O) and carbonaceous particles released by a traditional charcoal kiln were measured for 2 weeks. This controlled experiment allowed EFs for the dominant C and N species to be determined under field conditions. Charcoal kiln fumes are characterized, over the whole carbonization period, by very high average $\Delta CO/\Delta CO_2$ and $\Delta CH_4/\Delta CO_2$ ratios—24% and 6.8%, respectively (Table III). These ratios are approximately five times higher for CO and twenty times higher for CH_4 compared to the average molar ratio of savanna fires. This is another very clear illustration of the influence of burning conditions on the quality of volatile combustion products. The carbonization process, close to pyrolysis, favors the emission of reduced or incompletely oxidized compounds. For the other carbonaceous compounds (hydrocarbons, organic acids, and particulate C), as well as N compounds, the difference between the molar ratios from charcoal kilns and savanna fires is less important.

During the second Ivory Coast campaign in August 1994, a number of household fires using charcoal or firewood were studied. Charcoal combustion has very specific characteristics. It is a kind of glowing combustion that releases high amounts of CO ($\Delta CO/$

TABLE III

Average emission ratio (± standard deviation) measured over charcoal kilns and household wood and charcoal fires in Ivory Cost

$\Delta X/\Delta CO_2$ %	Charcoal making	Charcoal burning	Firewood burning
$\Delta CO/\Delta CO_2$	24.0±3.0	15.5±3.0	7.9±1.5
$\Delta CH_4/\Delta CO_2$	6.8±0.6	0.25±0.2	0.38±0.11
$\Delta NMHC/\Delta CO_2$	1.3±0.3	0.06±0.007	0.57±0.24
$\Delta OA/\Delta CO_2$	HCOOH=0.010±0.012	—	0.008±0.004
(organic acids)	CH_3COOH=0.16±0.14		0.065±0.034
$\Delta NO_x/\Delta CO_2$	0.02±0.01	—	0.15±0.04
$\Delta NH_3/\Delta CO_2$	0.05±0.06	—	—
$\Delta N_2O/\Delta CO_2$	0.0075±0.012	—	—

Source: Brocard *et al.*, 1995

ΔCO_2= 15.5%) but very little methane and NMHC. Firewood burning, at least using traditional West African stoves, appears to be a rather efficient flaming combustion very close to savanna burning; with similar ERs. The results from the second campaign, which are shown in Figure 6, are in accordance with average EFs for carbonaceous compounds emitted by wood-burning stoves given by Smith (1991). The Smith (1991) study is based on an extensive data set, including measurements of the energy budgets of traditional stoves.

The procedure for calculating EFs from molar emission ratios for woodburning is the same as the one suggested in "Emission Ratio and Emission Factor" in this article. However, for charcoal production, the C mass balance must be determined from direct measurements. This was done during the Ivory Coast campaign, leading to the estimated EFs shown in Table IV.

5.3. Forest Fires

Data on forest fires in tropical regions are very scarce because no large campaigns to determine all ERs and EFs from equatorial forest fires, such as SAFARI, have been organized until recently. However, a campaign (SCAR-C) is planned for the Amazon region in 1995.

Data come from preliminary experiments by Greenberg *et al.* (1984) or from the BASE-A (Kaufman *et al.*, 1992) and BASE-B (Ward *et al.*, 1992) experiments. The data primarily concern C compounds; they can be compared with those from chaparral forest or boreal forest fires (Cofer *et al.*, 1988; 1990) (Table V). The ranges of ERs for deforestation fires are very large and probably depend on sampling conditions. For such fires, helicopter or aircraft sampling seems to be the best approach. As compared to tropical

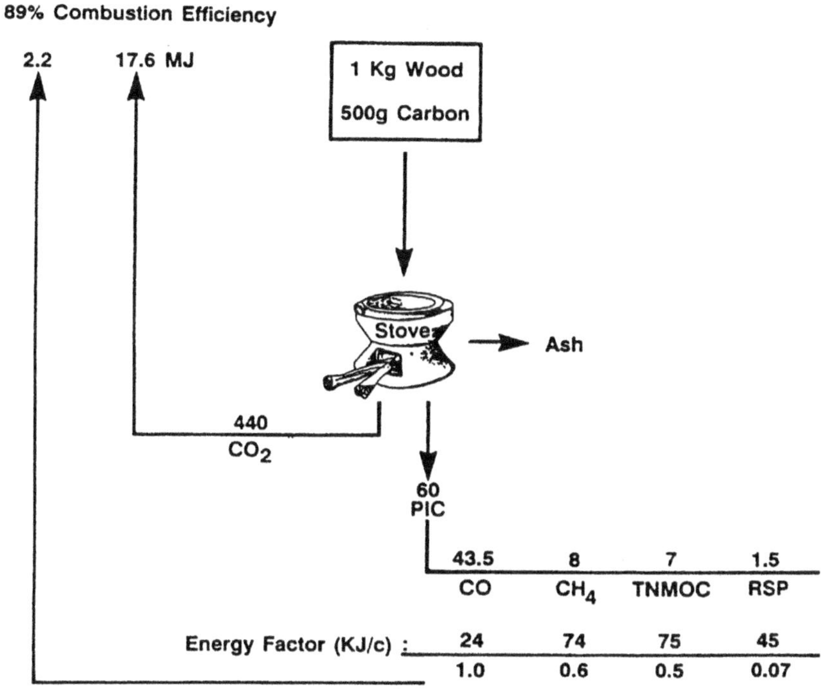

Figure 6. The cookstove energy cycle and carbon balance (from Smith, 1991).

forest fires, boreal fires have ERs close to that of savanna fires because of better flaming conditions. Tropical forest fires produce CO, CH_4, and NMHC ERs that are higher by a factor of two. Therefore, tropical forest fires are significant sources of these trace gases. EFs for tropical forest fires during BASE-B were calculated by Ward *et al.* (1992) using regression models that account for combustion efficiency (see "Emission Ratio and Emission Factor"). These EFs are presented in Table VI.

5.4. Agricultural Fires

Agricultural fires in the tropics include in situ burnings of any agricultural waste (for example, rice straw) and practices such as sugar cane burning before harvest. Of interest is whether the burning of agricultural residues (cereal straw, dung cakes) for domestic energy should be accounted for in the agricultural or residential sector. Because of the almost complete lack of data on EFs for this source, this issue is not important for this article; but it must be considered in any source estimate.

TABLE IV

Emission factors for biomass fuel burning determined from in situ measurements in Ivory Coast by Brocard *et al.* (1995)

	Charcoal making g C kg^{-1} dry wood	Firewood burning g C kg^{-1} dry wood	Charcoal burning g C kg^{-1} dry wood
$C(CO_2)$	120±25	400±65	170±30
C(C0)	30±8	30±10	25±9
$C(CH_4)$	8±2	1.5±0.6	0.5±0.4
C(NMHC)	2±0.6	2.5±1.2	0.1±0.03
C(O.A)	0.6	0.3	
		g N/kg dry wood	
$N(NO_X)$	0.02±0.02	0.69±0.18	—
$N(NH_3)$	0.07±0.10	—	—
$N(N_2O)$	0.02±0.04	—	—

During the SAFARI experiment, a series of samples was taken from aircraft over sugar cane fires and analyzed for CO_2, CO, CH_4, and NMHC concentrations (Koppmann *et al.*, 1995). Two different fires sampled in Swaziland produced significantly different ERs (Table VII). On average, these values were close to that of savanna fires or even lower, indicating very intense flaming conditions for sugar cane fires. The NO_2 measured in the smoke plume is not significant because NO_x is mainly emitted as NO.

TABLE V

Emission ratios (relative to CO_2) of trace gases and particles emitted by forest fires

Ecosystem	Reference	ER (%) CO	ER(%) CH_4	ER (%) NMHC	ER (%) H_2	ER (%) NO	ER (%) PM 2.5
Selva-Brazil (BASE-A)	Kaufmann et *al.*, 1992	5.4±2	0.3±0.1				0.6±0.30
Selva-Brazil (BASE-B)	Ward *et al.*, 1992	4.0-11.0	0.22-1.2	0.36-1.2	0.6-5.7	0.07-0.2	
Selva-Brazil	Greenberg *et al.*, 1984	11.9 4.7-30.2	1.1 0.5-2.3	1.2 0.7-2.0			
Boreal forest	Cofer III *et al.*, 1990	6.6±1.6	0.61±0.2	0.67±0.2	2.5±0.6		
Chapparal forest	Cofer III *et al.*, 1988	60±3.1	0.43±0.1	4.1±0.08	2.3±1.0		

TABLE VI

Emission factors for tropical fires as measured during BASE-A and BASE-B experiments.

Species	CO_2	CO	CH_4	NMHC	H_2	NO	PM 2.5
EF g/kg dry fuel	1 531–1 612	112–152	7.1–10.8	8–12	3.1–4.9	0.6–0.8	6.8–8.9

Nguyen *et al.* (1994) measured rice straw burning in Vietnam and estimated that 84% of the C burned is emitted as CO_2, 13% as CO, 1.1% as CH_4, and 1.1% as NMHC. Recent measurements of NO_x from cereal waste fires in Spain (Ezcurra *et al.*, 1995) give an average EF of 0.8 g(N)/kg dry matter.

6. Conclusion

Emission ratios and emission factors for savanna fires, the dominant form of biomass burning in the tropics, are well documented by important field campaigns conducted over the last 5 years. Two special issues on savanna fires should be published before the end of 1995: the FOS/DECAFE 1991 experiment (*Journal of Atmospheric Chemistry*) and the STARE-SAFARI 1992 experiment (*Journal of Geophysical Research*), as well as a book on the SAFARI experiment (Andreae *et al.* 1995a,b,c). In comparison, other types of biomass burning suffer from a serious lack of data. Nevertheless, for forest fires and biofuel combustion, the limited data sets available can be used to define realistic emission factors for global or country-by-country estimates of biomass burning emissions. There is also a dramatic lack of data on agricultural fires, which are important sources in many countries;

TABLE VII

Emission ratios of CO, CH_4, NMHC, and NO_2 measured over sugar cane fires in Swaziland during the SAFARI experiment.

Emission ratio to CO_2 (mol %)	CO	CH_4[a]	NMHC[a]	NO_2[b]
Sugar cane fire 1	4.4±2.1	0.26±0.2	0.39±0.29	0.047±0.006
Sugar cane fire 2	2.2±0.3	0.11±0.09	0.17±0.07	

[(a)]from Koppmann *et al.*, 1995
[(b)]from Harris *et al.*, 1995

however, default values provided by the well-documented studies on savanna fires can be used for carbonaceous compounds. For N compounds, the use of savanna fire data corrected with average values of N content in the plant biomass is suggested. For atmospheric trace gases and particles that impact the Earth's radiative balance, average emission factors for the dominant tropical burning processes are summarized in Table VIII. The results in Table VIII, which were taken from the most recent measurements, especially those in Africa, show good agreement with previous data used (for example, data in the VOC (volatile organic compound) estimate of the Global Emission Inventory Activity of IGAC) (Veldt, 1992). However, Table VIII is not complete, and data on forest fires and biofuel burnings are based on a very small number of experimental results. Further, there are no reliable data for agricultural burns. More research is needed to improve the data for all burning types other than savanna fires, especially agricultural fires.

Acknowledgment

The authors are very grateful to Prof. Dr. Meinrat O. Andreae for providing an advance overview paper that included results on emission factors measured during the SAFARI campaign. The SAFARI data set is the best one presently available and was extensively

TABLE VIII

Best-guess estimates of emission factors of greenhouse gases and aerosols from the main biomass burning sources in the tropics.

EF in g/kg dry fuel Species	Savanna fires	Forest fires	Biofuels Firewood g/kg dry wood	Biofuels Charcoal production g/kg dry wood	Charcoal burning g/kg charcoal
CO_2	1 640	1 580	1 460	440	2 240
CO	65	130	70	70	220
CH_4	2.4	9	2.0	10.7	2
NMHC (gC)	3.1	10	2.5	2.0	0.5
NO_X	3.1	0.7	3.7	0.11	
N_2O	0.15	—	0.15[a]	0.06	
CH_3Cl	0.11	—	0.11[a]		
PM 2.5 (gC)	5	8	3[a]	1	

Note: For emission calculations, the uncertainty on these average factors is estimated at ± 15% for savanna fires, based on the comparison of the various data sets, and ± 50% for the other sources due to limited data.

[a]Estimates from the other combustion types

used in this paper. We would also like to thank all the scientists involved in the SAFARI experiment, with a special mention to Dr. Joel S. Levine, who sent us in advance the SAFARI paper of his group and helped organize the last two Chapman Conferences on Global Biomass Burning.

References

Andreae, M.O.: 1991, Biomass burning: Its history, use and distribution, and its impact on environmental quality and global climate, in Levine, J.S. (ed.), *Global Biomass Burning,* MIT Press, 3–28.

Andreae, M.O., Chapuis, A., Cros, B., Fontan, J., Helas, G., Justice, C., Kaufman, Y.J., Minga, A., and Nganga, D.: 1992, Ozone and Aitken nuclei over equatorial Africa: airborne observations during DECAFE 88, *Journal of Geophysical Research,* **97,** 6137–6148.

Andreae, M.O., Atlas, E., Cachier, H., Coffer III, W.R., Harris, G.W., Helas, G., Koppmann, R., Lacaux, J.P., and Ward, D.E.: 1995b, Trace gas and aerosol emissions from savanna fires, in Levine, J.S., (ed.), *Biomass Burning and Global Change,* MIT Press, submitted for publication.

Andreae, M.O., Andreae, T.W., Annegarn, H., Beer, F., Cachier, H., Helbert, W., Harris, G.W., Maenhaut, W., Salma, I., Swap, R., Wienhold, F.G., and Zenker, T.: 1995a, Airborne studies of aerosol emissions from savanna fires in southern Africa 2: Aerosol chemical composition, *Journal of Geophysical Research,* in press.

Andreae, M.O., Atlas, E., Harris, G.W., Helas, G., de Kock, A., Koppmann, R., Maenhaut, W., Manö, S., Pollock, W.H., Rudolph, J., Scharffe, D., Schebeske, G., and Welling, M.: 1995c, Methyl halide emissions from savanna fires in southern Africa, *Journal of Geophysical Research,* in press.

Ahuja, D.: 1994, Methane emissions from anthropogenic biomass burning, in *International Anthropogenic Methane Emissions: Estimates for 1990,* EPA Report 230–R–93–010, chapter 4.

Atlas, E., Pollock, W., and DeCock, A.: 1995, Organic compound emissions from biomass burning during the SAFARI experiment, *Journal of Geophysical Research,* in press.

Ayers, G.P., and Gillett, R.W.: 1988, Isoprene emissions from vegetation and hydrocarbon emissions from bushfires in tropical Australia, *Journal of Atmospheric Chemistry,* **7,** 177–190.

Bonsang, B., Lambert, G., and Boissard, C.: 1991, Light hydrocarbon emissions from African savanna burnings, in Levine, J.S. (ed.), *Global Biomass Burning,* MIT Press, 155–161.

Bonsang, B., Boissard, C., Le Cloarec, M.F., Rudolph, J., and Lacaux, J.P.: 1995, Methane, carbon monoxide and light nonmethane hydrocarbon emissions from African savanna burnings during the FOS/DECAFE experiment, *Journal of Atmospheric Chemistry,* in press.

Brocard D., Lacaux, C., Lacaux, J.P., Kouadio, G., Yoboué, V., Assa Achy, M., Ahoua, B., and Koffi, M.: 1995, Emissions from the combustion of biofuels in the African tropics, in Levine, J.S. (ed.), *Biomass Burning and Global Change,* MIT Press, to be published.

Brown, S., Hall, C.A., Knabe, W., Raich, J., Trexler, M.C., and Woomer, P.: 1993, Tropical forests: Their past, present and future role in terrestrial carbon budget, *Water, Air and Soil Pollution,* **70,** 71–94

Cachier H., Ducret, J., Brémond, M., Yoboué, V., Lacaux, J.P., Gaudichet, A., and Baudet, J.: 1991, Biomass burning in a savanna region of the Ivory Coast, in Levine, J.S. (ed.), *Global Biomass Burning: Atmospheric, Climatic and Biospheric Implications,* MIT Press, 174–180.

Cachier, H.: 1992, Biomass burning sources, in *Encyclopedia of the Earth System Science,* **1,** Academic Press.

Cachier H., Liousse, C., Buat-Menard, P., and Gaudichet, A.: 1995, Particulate content of savanna fire emissions, *Journal of Atmospheric Chemistry,* in press.

Clain, M.P.: 1995, Etude de la composante atmosphérique de l'aérosol organique, Ph.D. thesis, Université de Savoie, Chambery, France.

Cofer III, W.R., Levine, J.S., Riggan, J., Sebacher, D.I., Winstead, E.L., Shaw, Jr., E.F., Brass, J.A., and Ambrosia, V.G.: 1988, Trace gas emissions from a mid-latitude prescribed chaparral fire, *Journal of Geophysical Research,* **93,** 1653–1658.

Cofer III, W.R., Levine, J.S., Winstead, E.L., and Stocks, B.J.: 1990, Gaseous emission from Canadian boreal forest fire, *Atmosphere and Environment,* **24,** 1653–1659.

Cofer III, W.R., Levine, J.S., Winstead, E.L., Cahoon, D.R., Sebacher, D.I., Pinto J.P., and Stocks, B.J.: 1995, Source compositions of trace gases released during African savanna fires, in Levine, J.S. (ed.), *Biomass Burning and Global Change*, MIT Press.

Cros, B., Delmas, R., Nganga, D., Clairac, B., and Fontan, J.: 1988, Seasonal trends of ozone in equatorial Africa: experimental evidence of photochemical formation, *Journal of Geophysical Research*, **93**, 8355–8366.

Crutzen, P.J., Heidt, L.E., Krasneck, J.P., Pollock, W.H., and Seiler, W.: 1979, Biomass burning as a source of atmospheric trace gases CO, H_2, N_2O, NO, CH_3Cl, and COS, *Nature*, **282**, 253–256.

Crutzen, P.J., Delany, A.C., Greenberg, J., Haagenson, P., Heidt, L., Lueb, R., Pollock, W., Seiler, W., Wartburg, A., and Zimmerman, P.: 1985, Tropospheric chemical composition measurements in Brazil during the dry season, *Journal of Atmospheric Chemistry*, **2**, 233–256.

Crutzen, P.J., and Andreae, M.O.: 1990, Biomass burning in the tropics: Impact on atmospheric chemistry and biogeochemical cycles, *Science*, **250**, 1669–1678.

Crutzen, P.J., and Goldammer, J.G. (eds.): 1995, *Fire in the Environment: The Ecological, Atmospheric, and Climatic Importance of Vegetation Fires*, John Wiley and Sons Ltd.

Delmas, R.: 1982, On the emission of carbon, nitrogen and sulfur in the atmosphere during savanna fires in intertropical savanna zones, *Geophysical Research Letters* **9**, 761–764.

Delmas, R.A., Marenco, A., Tathy, J.P., Cros, B., and Baudet, J.G.R.: 1991a, Sources and sinks of methane in African savanna. Methane emission from biomass burning, *Journal of Geophysical Research*, **96**, 7287–7299.

Delmas, R.A., Loudjani, Ph., Podaire, A., and Menaut, J.C.: 1991b, Biomass burning in Africa: an assessment of annually burned biomass, in Levine, J.S. (ed.), *Global Biomass Burning*, MIT Press, 126–132.

Delmas R.A.: 1994, An overview of present knowledge on methane emission from biomass burning, *Fertilizer Research*, **37**, 181–190.

Delmas R.A., Lacaux, J.P., Menaut, J.C., Abbadie, L., Le Roux, X., Helas, G., and Lobert, G.: 1995, Nitrogen compound emission from biomass burning in tropical African savanna, FOS/DECAFE 91 Experiment (Lamto, Ivory Coast), *Journal of Atmospheric Chemistry*, in press.

Dixon, R.K., Brown, S., Houghton, R.A., Solomon, A.M., Trexler, M.C., and Wisniewski, J.: 1994, Carbon pool and flux of global forest system, *Science*, **263**, 185–190.

Ezcurra, A., de Zarate, O., Lacaux, J.P., and Dinh, P.V.: 1995, Emission factors of cereal waste burning, 2nd Conference on Global Biomass Burning, Williamsburg, VA, March.

Fishman J., Watson, C.E., Larsen, J.C., and Logan, J.A.: 1990, Distribution of tropospheric ozone determined from satellite data, *Journal of Geophysical Research*, **95**, 3599–3617.

Fishman J., Fakhruzzaman, K., Cros, B., and Nganga, D.: 1991, Identification of widespread pollution in the Southern Hemisphere deduced from satellite analyses, *Science*, **252**, 1693–1696.

Gaudichet, A., Echalar, F., Chatenet, B., Quisefit, J.P., Malingre, G., Cachier, H., Buat-Menard, P., Artaxo, P., and Maenhaut, W.: 1995, Trace elements in tropical African savanna biomass burning aerosols, *Journal of Atmospheric Chemistry*, in press.

Greenberg, J.P., Zimmerman, P.R., Heidt, L., and Pollock, W.: 1984, Hydrocarbon and carbon monoxide emissions from biomass burning in Brazil, *Journal of Geophysical Research*, **89**, 1350–1354.

Griffith, D.W.T., Mankin, W.G., Coffey, M.T., Ward, D.E., and Riebau, A.: 1991, FTIR remote sensing of biomass burning emissions of CO_2, CO, CH_4, NO, NO_2, NH_3, and N_2O, in Levine, J.S. (ed.), *Global Biomass Burning*, MIT Press, 230–239.

Goldammer, J.G. (ed.): 1990, *Fire in the Tropical Biota*, Springer-Verlag.

Hao, W.M., Liu, M.H., and Crutzen, P.J.: 1990, Estimates of annual and regional releases of CO_2 and other trace gases to the atmosphere from fires in the tropics, based on the FAO statistics for the period 1975-1980, in Goldammer, J.G. (ed.), *Fire in the Tropical Biota*, Springer-Verlag.

Hao, W.M., and Ward, D.E: 1993, Methane production from global biomass burning, *Journal of Geophysical Research*, **98**, 20,657–20,661.

Harris, G.W., Zenker, T., Wienhold, F.G., Welling, M., Parchatka, U., and Andreae, M.O.: 1995, Airborne measurements of trace gas emission ratios from southern African veld fires, *Journal of Geophysical Research*, in press.

Harris, G.W., Zenker, T., Wienhold, F.G., Welling, M., and Parchatka, U.: 1993, Airborne measurements of trace gas emission ratios from southern African veld fires, *EOS*, **74**, 104.

Hegg, D.A., Radke, L.F., Hobb, P.V., Rasmussen, R.A., and Riggan, P.J.: 1990, Emissions of some trace gases from biomass fires, *Journal of Geophysical Research*, **95**, 5669–5675.

Helas, G., Andreae, M.O., Schafer, L., Goldammer, J., Scharffe, D., Lobert, J., Ajavon, A., Brou, A., Lacaux, J.P., and Delmas, R.: 1995, Atmospheric chemical aspects of biomass burning/Results from savanna fires in Cote d'Ivoire, January 1991, DECAFE 91/FOS, *Journal of Atmospheric Chemistry*, in press.
Hurst, D.F., Griffith, D.W.T., Carras, J.N., Williams, D.J., and Fraser, P.J.: 1994a, Measurements of trace gases emitted by Australian savanna fires during the 1990 dry season, *Journal of Atmospheric Chemistry*, **18**, 33–56.
Hurst, D.F., Griffith, D.W.T., and Cook, G.D.: 1994b, Trace gas emissions from biomass burning in tropical Australian savannas, *Journal of Geophysical Research*, **99**, 16,441–16,456
Kaufman, Y.J., Setzer, A., Ward, D., Tanre, D., Holben, B.N., Menzel, P., Pereira, M.C., and Rasmussen, R.A.: 1992, Biomass burning airborne and spaceborne experiment in the Amazonas (BASE-A), *Journal of Geophysical Research*, **97**, 14,581–14,599.
Kirchhoff, V.W.J.H., Setzer, A.W., and Pereira, M.C.: 1989, Biomass burning in Amazonia: Seasonal effects on atmospheric O_3 and CO, *Geophysical Research Letters*, **16**, 469–472.
Koppmann, R., Khedim, A., Rudolph, J., Helas, G., Welling, M., and Mano, S.: 1995, Airborne measurements of nonmethane hydrocarbon emissions from savanna fires in southern Africa, *Journal of Geophysical Research*, in press.
Kuhlbusch, T.A., Lobert, J.M., Crutzen, P.J., and Warneck, P.: 1991, Molecular nitrogen emissions from denitrification during biomass burning, *Nature*, **351**, 135–137.
Lacaux, J.P., Cachier, H., and Delmas, R.: 1993, Biomass burning in Africa: An overview of the impact on the atmospheric chemistry, in Crutzen P.J., and Goldammer, J.G. (eds.), *Fire in the Environment: The Ecological, Atmospheric, and Climatic Importance of Vegetation Fires*, John Wiley and Sons Ltd., 160–191.
Lacaux J.P., Brocard, D., Lacaux, C., Delmas, R., Ahoua, B., Yoboué, V., and Koffi, M.: 1994, Traditional charcoal making: An important source of atmospheric pollution in the African tropics, *Journal of Atmospheric Research*, **35**, 71–76.
Lacaux, J.P., Brustet, J.M., Delmas, R., Menaut, J.C., Abbadie, L., Bonsang, B., Cachier, H., Baudet, J., Andreae, M.O., and Helas, G.: 1995, Biomass burning in the tropical savannas of Ivory Coast: An overview of the field experiment Fire of Savannas (FOS/DECAFE 91), *Journal of Atmospheric Chemistry*, in press.
Le Canut, P., Andreae, M.O., Harris, G.W., Wienhold, F.G., and Zenker, T.: 1995, Airborne studies of aerosol emissions from savanna fires in southern Africa: Results from the laser-optical particle counter, *Journal of Geophysical Research*, submitted for publication.
Levine, J.S. (ed.): 1991, *Global Biomass Burning: Atmospheric, Climatic and Biospheric Implications*, MIT Press.
Levine J.S., Cofer III, W.R., Winstead, E.L., Rhinehart, P., Cahoon, D.R., Sebacher, D.I., Sebacher, S., and Stocks, B.B.J.: 1991, Biomass burning: Combustion emissions, satellite imagery and biogenic emissions, in Levine, J.S. (ed.), *Global Biomass Burning*, MIT Press, 264–271.
Lobert, J.M., Scharffe, D.H., Hao, W.M., Kuhlbusch, T.A., Seuwen, R., Warneck, P., and Crutzen, P.J.: 1991, Experimental evaluation of biomass burning emissions: nitrogen carbon containing compounds, in Levine, J.S. (ed.), *Global Biomass Burning*, MIT Press, 289-304.
Maenhaut, W., Salma, I., Cafmeyer, J., Annegarn, H.J., and Andreae, M.O.: 1995, Regional atmospheric aerosol and sources in the eastern Transvaal, South Africa, and impact of biomass burning, *Journal of Geophysical Research*, in press.
Marenco, A., Medale, J., and Prieur, S.: 1990, Study of tropospheric ozone in the tropical belt (Africa-America) from STRATOZ and TROPOZ campaigns, *Atmosphere* and *Environment* **24A**, 2823–2834.
Marenco, A., Prieur, S., Nedelec, P., Macaigne, M., and Volz, G.: 1992, Large scale study of CO and CH_4 in the background troposphere from meridional airborne campaigns TROPOZ II and SRTATOZ III, SVII EGS General Assembly, Edinburgh, Scotland.
Nguyen, B.C., Mihalopoulos, N., and Putaud, J.P.: 1994, Rice straw burning in southeast Asia as a source of CO and COS to the atmosphere, *Journal of Geophysical Research*, **99**, 16,435–16,439.
Seiler, W., and Crutzen, P.J.: 1980, Estimates of gross and net fluxes of carbon between the biosphere and the atmosphere from biomass burning. *Climate Change*, **2**, 204-247.
Smith, K.R.: 1991, Biomass cookstoves in global perspectives: energy, health and global warming, in *The Earth as System Central*, project main report 3, EWC/ESMAP/UNDP, evaluation of improved cookstove programs, World Bank, Washington, DC.
Veldt, C.: 1992, GEIA note on residential biomass burning, IGAC.

Ward, D.E., Susott, R.A., Kauffman, J.B., Rabbitt, R.E., Cummings, D.L., Dias, B., Holben, B.N., Kaufman, Y.J., Rasmussen, R.A., and Setzer, A.W.: 1992, Smoke and fire characteristics for cerrado and deforestation burns in Brazil: BASE-B experiment, *Journal of Geophysical Research,* **97,** 14,601–14,619.
Ward, D.E., and Radke, L.F.: 1993, Emission measurements from vegetation fires: A comparative evaluation of methods and results, in Crutzen, P.J., and Goldammer, J.G. (eds.), *Fire in the Environment: The Ecological, Atmospheric, and Climatic Importance of Vegetation Fires,* John Wiley and Sons Ltd., 53–76.

DETERMINING THE GLOBAL SIGNIFICANCE OF LOCAL AND REGIONAL MITIGATION STRATEGIES: SETTING THE SCENE WITH GLOBAL INTEGRATED ASSESSMENT MODELS

R. LEEMANS

Department of Terrestrial Ecology and Global Change, National Institute of Public Health and Environmental Protection, RIVM, PO Box 1, 3720 BA Bilthoven, The Netherlands

Abstract. Because of the importance and complexity of the processes involved in the response of land cover and land use to changing environmental conditions, other approaches are required to evaluate the effectiveness of mitigation options. One such approach is provided by the global integrated assessment model, IMAGE 2. This article presents the structure and some of the underlying assumptions of IMAGE 2, which illustrates the importance of feedback processes in evaluating the effectiveness of mitigation options with respect to land use. Although models such as IMAGE 2 are unsuitable for local and national mitigation evaluations, they can be used to define the regional and global constraints of the smaller scale assessments.

1. Introduction

National emission inventories of greenhouse gases (GHGs) are an important first step in determining the role of regional emissions in the buildup of atmospheric concentrations of GHGs. Adequate quantification of this buildup could enhance understanding of the local, regional, and global biogeochemical and physical processes involved—but only if smaller scale studies are based on compatible methodologies that allow them to be integrated into broader studies with a continental and even global scope.

The International Panel on Climate Change (IPCC) methodology for estimating national emissions (UNEP *et al.*, 1995) is appropriate for assessing the current situation. The IPCC approach focuses on calculating the fluxes of the major GHGs from different sectors. Straightforward conversion functions are applied, for example, to compute total annual carbon (C) storage in forests by making use of their areal extent and annual biomass and C accumulation rates under different management conditions. The functions required are well established for current conditions and can be implemented from different types of sectoral records, such as forest inventories, but the data needed to do so are not always readily available. However, under changed atmospheric conditions and climate, these conversion functions (especially accumulation rates) could change considerably. This potential for change requires careful validation and renewed calibration when applying them repeatedly, or under other conditions. This article addresses this issue in more detail and attempts to establish a systematic framework in which these functions change as a consequence of environmental change. Such a framework allows for a more comprehensive local and regional inventory of GHG emissions and evaluation of different mitigation potentials.

Environmental Monitoring and Assessment **38**: 205-216, 1995.

The objective of this article is to highlight some of the cross-sectoral issues and integrating factors between emission inventories and mitigation options. Discussed first are some of the processes involved and their interactions within the Earth system (climate, oceans, biosphere, and society). Because of the complexity of this system, discussion is limited to processes involving the biosphere, human society, and the atmosphere (including climate), with emphasis on land-use and land-cover change. Implications of these systemic interactions are then illustrated by presenting the characteristics and some simulations from a global integrated assessment model. Such a model does not simulate the full complexity of the Earth's system, but rather is especially designed to project future atmospheric concentrations of GHGs for different societal and environmental assumptions. Although there are now several such models available (Morita *et al.*, 1993; Edmonds *et al.*, 1994), only the IMAGE 2 model (Alcamo, 1994) is discussed, as it is currently the most advanced with respect to issues involving land-use and land-cover change. Although IMAGE 2 addresses the dynamics of all GHGs, this article mainly focuses on carbon dioxide (CO_2).

2. Significance of land-use and land-cover change

Land cover differs from land use (Turner *et al.*, 1993). "Land" refers to the relatively stable solid surface of the Earth, including the soil with its underlying geology and hydrology, and the biosphere and atmosphere. "Land cover" refers to the actual appearance of the land and is often understood as vegetation or ecosystems. "Land use" refers to the purpose for which humans exploit land and land cover. Land use consists, therefore, of a series of management activities that alter land cover. The consequences of land use and land-use change for land cover are obvious and relevant to environmental change. Land-cover characteristics characterize ecosystems; together, they define the interface between the atmosphere, climate system, biosphere, and society.

These definitions are broader than those given in the IPCC Inventory Guidelines (UNEP *et al.*, 1995), where emissions from land-use change are mainly those characterized by conversions of different land-cover types. In the IPCC approach one land-cover type changes into another (for example, from forests into grasslands). However, land-use change results in two types of change to land and land cover: conversions and modifications (cf. Turner *et al.*, 1990; Leemans and Zuidema, 1995). Modification of a specific land cover does not change its overall class, but rather its properties. Many agricultural and forestry practices are good examples of continuous modification.

Both types of change are important to environmental change, because they affect water availability by altering surface characteristics (roughness, albedo) and the dynamics of the biogeochemical cycles (C, N, and H_2O). As a consequence, these changes can have severe impacts on soil quality, ecosystem productivity, and biodiversity. For example, as a result of land-cover change, C could be rapidly lost from areas where forests die because of drought, and climate could be changed locally because of changes in boundary layer characteristics, resulting in largely different patterns of C storage. Such responses result from the complex feedback interactions within a highly dynamic system. The response of such processes can be determined by inventories only if they are repeated regularly; however, future responses can never be quantified by extrapolating trends of repeated inventories because of the nonlinearities involved in the processes.

A feedback is a response in which the factor causing an effect is modified by that effect. A feedback is positive if the response increases, or negative if the response decreases. If feedback responses are important components of a system, a simple cause-process-effect modeling or budget approach will hamper a precise quantification of the effect. The conclusions derived from such assessments can be misleading, especially if the results are extrapolated to future conditions.

The response of the biosphere to changes in land cover, atmospheric composition, and climate is highly complex. Several direct and indirect responses can be recognized (Figure 1). For example, C storage in ecosystems probably decreases in a warmer climate as a direct result of the lower responsiveness of photosynthesis to total plant and soil respiration. However, C storage could increase in woody ecosystems through the indirect effect of nutrient release by an increase in decay rate. Such feedback processes influence the functioning of ecosystems and, besides producing changes in C storage, also shift land-cover patterns and the composition and structure of

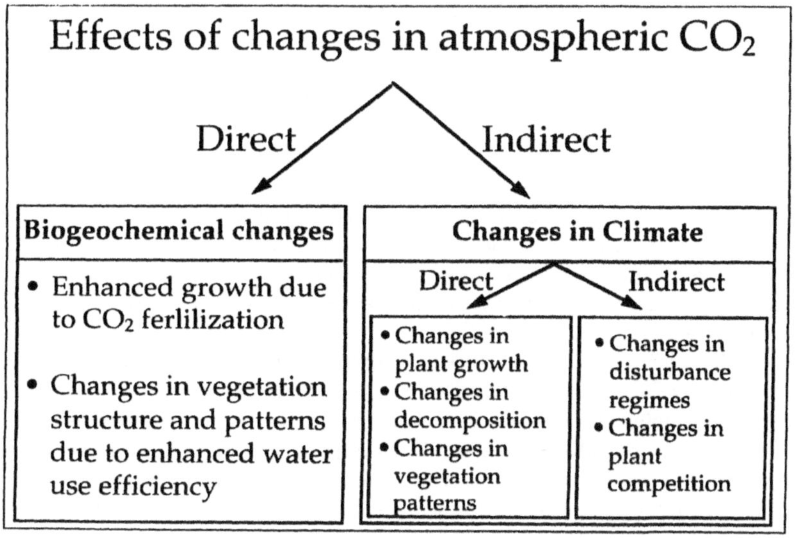

Figure 1. Effects of changes in atmospheric CO_2 on terrestrial systems and biogeochemical cycles.

ecosystems (Cramer and Leemans, 1993; Jackson *et al.*, 1994; Schulze and Mooney, 1993).

Several feedback processes could lead to increased C storage. For example, shifts in vegetation zones under a warmer climate could increase total C storage (Smith *et al.*, 1992). Such changes in global C storage have been observed in the past (Van Campo *et al.*, 1993). However, these analyses show large local and regional differences (Kirschbaum, 1993), which are not only a result of different responses of different land-cover types, but also a consequence of local and regional changes in the causal factors, land-cover change, atmospheric composition, and climate. Consequently, emission inventories and assessments of mitigation potential should account for such spatial heterogeneity in land-cover characteristics and response patterns.

3. IMAGE 2: an integrated assessment model

3.1. Short Description of the Model

The first version of IMAGE (Rotmans *et al.*, 1990) was a globally aggregated model that computed the atmospheric GHG emissions from energy use, industrial activities, and land use (that is, deforestation). Final concentrations were computed while accounting for simple chemical processes in the atmosphere and C uptake elsewhere in the global C cycle. The final radiative forcing of these GHGs was determined and used by IMAGE to compute global annual mean temperature changes and sea-level rises. The impacts of such changes were scarcely considered or not considered at all, let alone the consequences of important feedbacks and the spatial heterogeneity of land-use and land-cover change.

IMAGE 2 was developed to overcome these limitations. Details on the underlying assumptions, implementation, and structure of the model are given by Alcamo (1994). This article only summarizes some of IMAGE 2's main features.

The model is designed to provide a dynamic and long-term overview of climate-change issues to support a scientific evaluation of national and international policies on the buildup of GHGs in the atmosphere. To provide reliable future estimates, the model incorporates the different linkages and feedbacks and their relative importance in the society-biosphere-climate system.

IMAGE 2 consists of a set of linked and simplified submodels, each of which has been tested and validated against available data sets. The linkages between the submodels consist of transparent information flows that allow for tests on uncertainties and implementation of different assumptions. Most socioeconomic causal input variables, such as per capita income, population densities, fuel mixes, and technology levels, are provided to IMAGE 2 as boundary conditions. This allows for objective analyses of the consequences of different trends in these variables.

IMAGE 2 consists of three fully linked components (Figure 2): the Energy-Industry System (EIS), the Terrestrial Environment System (TES), and the Ocean-Atmosphere System (OAS). Dynamic calculations are performed for a 100-year time horizon. EIS and the land-use component of TES are regionally explicit (that is, thirteen "homogeneous" countries, sets of countries, or continents), while all environmental and physical submodels are geographically explicit. The land-cover models of TES are implemented on a terrestrial grid of 0.5° longitude and latitude. Each grid cell is characterized by its specific climate, vegetation, and soil properties. The atmosphere in OAS is subdivided into 10° latitudinal belts and several vertical layers; the ocean is simulated as a series of different boxes, reproducing the different processes for each part of the ocean.

The objective of EIS is to compute the emissions of GHGs in each region as a function of energy consumption and industrial production. The models are designed especially for investigating the effectiveness of changed fuel mixes, improved energy efficiency, and technological development on regional GHG emissions.

The objective of TES is to simulate global land use and land cover dynamically through time and employ land use and changes in land-cover patterns to compute GHG fluxes between the biosphere and the atmosphere. Land use is defined by demand for commodities, such as crops, livestock, lumber, and fuelwood, and is determined for each region on the basis of characteristic dietary preferences, income levels, and population densities. Land-use potential for each cell is determined from its characteristic rain-fed crop productivity and technology and management levels; land cover is defined by climate and soil characteristics. Intensification of agricultural land is explicitly considered. Changes in land cover are simulated by applying heuristic rules reconciling the current land-cover patterns with changes in demand. The rules reflect the most obvious controls, such as proximity to current infrastructure, population, and the most productive lands.

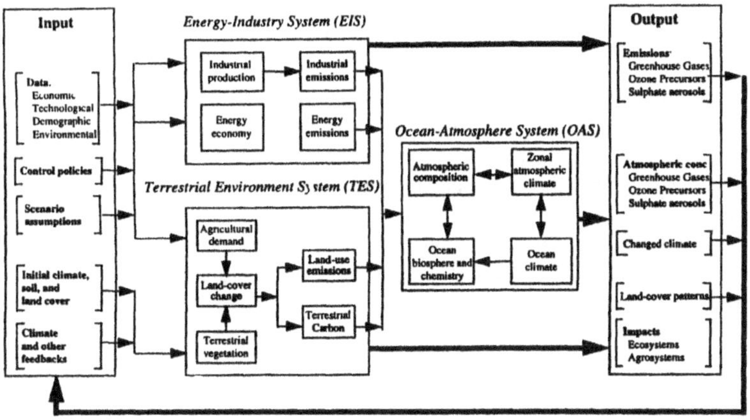

Figure 2. Conceptual diagram of the IMAGE 2 model.

IMAGE 2 simulates a dynamically changing land cover. Increased demand for land does not automatically lead to deforestation. All land-cover types with a potential for agricultural production can be converted into new agricultural or grazing lands. The heuristic model generates regionally different patterns of land-use change. In several developed regions the extent of agricultural land decreases, while in others there is a large increase. These dynamics are a result of the regional differences in driving forces.

Simulations of the baseline scenario, which has strong similarities to the IPCC IS92a scenario (Pepper *et al.*, 1992), reveal large regional shifts in reforestation and deforestation patterns: Agricultural areas contract in the more developed regions as defined in IMAGE 2, which leads to increased forestation and C storage. Agricultural expansion occurs in most other regions of the world, but gradually declines in the some of these regions. At the end of the simulations, there is a stable land cover in the India and China regions, while a slight contraction has occurred in the Latin America region. The simulations also show shifts of the current agricultural zones due to climate change. The resulting land-cover and agricultural activities serve as a basis for computing non-CO_2 GHG emissions. Each grid cell is further characterized by its own C cycle, with its specific land-use and environment-dependent C stocks and fluxes. Converting the grid cell's original land cover to agriculture produces a significant CO_2 emission from its total C stock. Feedbacks caused by changes in atmospheric CO_2 concentrations and climate are accounted for.

The purpose of OAS is to compute the buildup of GHGs in the atmosphere and the resulting change in temperature and precipitation patterns. Emissions from EIS and TES are combined and used to determine the physical and biological oceanic C uptake. The atmospheric composition model simulates globally averaged tropospheric and stratospheric chemistry, including some of the processes related to ozone production and destruction. The atmospheric energy balance submodel combines the resulting atmospheric levels of GHGs with additional data to compute the Earth's energy balance and climate for latitudinally averaged zones. The longitudinal dimension is obtained by scaling these zonal averages to the terrestrial grid using results from global atmospheric circulation models.

TES is linked to EIS via the demand for fuelwood and modern biomass for energy generation; TES also adequately addresses the land requirements for this demand. TES and OAS are directly linked through the GHG emissions, and indirectly through changes in soil moisture and surface characteristics caused by land-cover change. OAS is linked with TES through its climatic change patterns.

The whole IMAGE 2 framework provides an integrated simulation of important compartments of the Earth's system, allowing for an adequate integrated assessment of different mitigation options and strategies. However, data availability and quality do not yet allow for precise regional assessments. The improved land-cover data sets obtained by remote sensing and the basic data collected by the different national emission inventories can help improve the simulations.

3.2. Defining the Significance of Feedback Processes

The C cycle model implemented in TES can evaluate the importance of different feedback processes. Simulations show that each feedback has different regional characteristics (Van Minnen *et al.*, in press). Despite the fact that the atmospheric CO_2 concentration is homogeneously distributed, the additional C sequestration of enhanced plant growth under higher CO_2 concentrations (the so-called CO_2 fertilization effect) is most obvious in the forested tropical ecosystems. This is because the other tropical ecosystems have a large component of C4 species, which do not show strong responses, and because temperate and boreal plants grow under a lower temperature domain, which hampers the fertilization effect. CO_2 fertilization represents a negative feedback because it reduces the buildup of the atmospheric CO_2 concentration by sequestering more C in the vegetation.

The influence of climate change (both temperature and moisture availability) enhances plant growth throughout the world, but is most pronounced in boreal ecosystems because the plants there are the most highly temperature-limited ones. Also, the highest temperature increases are projected in these high-latitude regions. Globally, this process results in a stronger negative feedback than the CO_2 fertilization feedback. The influence of climate change on soil decomposition reduces the C release somewhat in the tropics, so that more C will be sequestered in tropical soils. The influence of climate change results in generally dryer tropical soils, while in the mid- and high latitudes decomposition accelerates and results in a strong positive feedback. If all the different feedback processes are combined, the highly nonlinear character of the response becomes obvious. The additional C sequestration for the combined feedback cannot be predicted from the individual feedback processes because of nonlinear response functions. Initially, the combined feedback processes result in enhanced sequestration, but this effect declines with a continuing increase of the CO_2 concentration and warming.

3.3. Evaluating Mitigation Potential

One of the major mitigation options given in many assessments is the adoption of renewable energy sources, such as the use and cultivation of modern biomass for energy generation (Johansson *et al.*, 1993). Although feasible and probably highly recommendable, this option encompasses a novel type of land use and could thus have great consequences for land-cover patterns and the extent of agricultural areas.

To investigate the sensitivity of different energy assumptions on land use and land cover, the baseline scenario (Figure 3) has been changed slightly. In this scenario, modern biofuels satisfy 75 EJ and 210 EJ of the total energy production in 2050 and 2100, respectively. The baseline has been based on the assumption that this requirement can be easily met by using crop residues and different kinds of wastes (Johansson *et al.*, 1993). Biofuels therefore do not require additional agricultural land.

Two other scenarios were run with variations in this assumption. One, the biofuel crop scenario, assumes that residues cannot adequately satisfy this total biofuel demand and that 40% of the energy comes from modern biomass crops with short rotations, such as

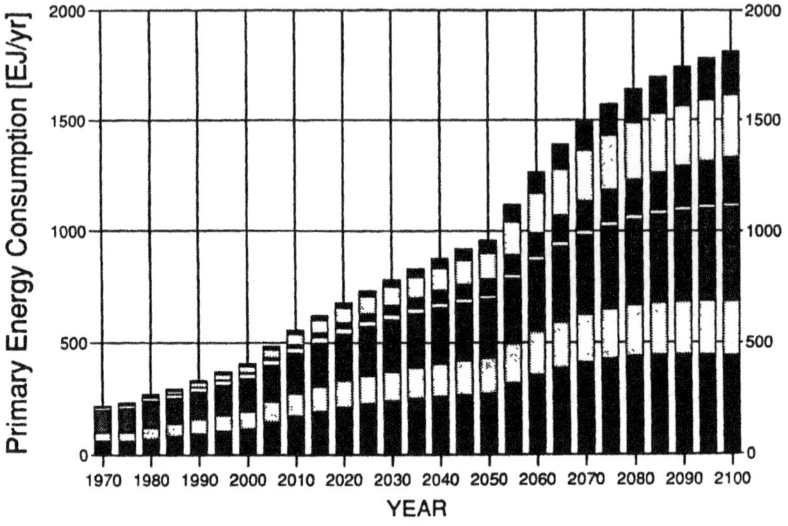

Figure 3. Total primary energy consumption (EJ a^{-1}; 1 EJ = 10^{18} J) by energy carriers for the baseline scenario of IMAGE 2 (the verticals bars present from bottom to top: coal, natural gas, oil, fuelwood, biomass, nuclear and renewable energy carriers).

Salix sp. and *Miscanthus* sp. The third scenario assumes that no biofuels are used at all and that energy demand is satisfied by fossil oil.

Table I lists the consequences of these three scenarios for the amount of global agricultural land needed as well as global deforestation and atmospheric CO_2. The biofuel crop scenario requires the largest amounts of new agricultural lands, which cannot be used for durable C sequestration. The atmospheric CO_2 concentration is therefore still considerably higher than in the baseline scenario, and only a little lower than in the oil scenario. Note, however, that the deforestation rates are only slightly higher in the biofuel scenario than in the baseline and the oil scenarios: the heuristic rule-based land-use change model projects large biofuel plantations on grasslands and other nonforested lands.

An unexpected result of this analysis was the effect on emissions of other GHGs, especially methane, nitrous oxide, and the ozone precursors (Alcamo *et al.*, 1994). The baseline and biofuels scenarios gave higher concentrations of these gases, mainly through burning of residues and intensified agricultural activities. The total radiative forcing of the baseline scenario and the oil scenario did not differ significantly, while the biofuel crop scenario showed greater climatic impacts. Mitigation measures aimed at reducing only a single GHG, therefore, may not be the most effective. The emissions of all GHGs must

TABLE I

Difference in atmosphere CO_2 concentration and land cover patterns for the different assumptions implemented to satisfy future energy demand

	Atmospheric CO_2 Concentration (ppm)	Agricultural area (10^3 km^2)	Forest area (10^3 km^2)
1990	492	26.7	47.2
2050			
Baseline	522	+9%	–26%
Biofuel crops	534	+30%	–32%
Oil	539	+9%	–26%
2100			
Baseline	777	+14%	–27%
Biofuel crops	821	+65%	–31%
Oil	857	+15%	–27%

Additional energy needs (above 1990 levels): 75 EJ in 2050 and 210 EJ in 2100. Assumptions: Baseline: 100 percent crop residues; Biofuel crops: 60 percent crop residues and 40 percent modern biomass crops; Oil: 100 percent fossil oil and no biofuels.

be evaluated in determining the relative effectiveness of abatement policies. Above all, land use and the resulting land-cover patterns should play a major role in such evaluations.

IMAGE 2 has also been used to evaluate land requirements for future global mitigation potential through forestation. New forests store C, and if these forests are left undisturbed or their wood products are used in durable applications, additional C storage could be considerable. Using forest products and their residues as biofuels could further decrease dependence on fossil fuels. Such GHG abatement policies have received considerable positive attention (Winjum *et al.*, 1993), but have not yet been able to offset significant CO_2 emissions from fossil fuels (Sedjo and Solomon, 1989). However, because forests satisfy many other functions, such as fuelwood supply, erosion management, recreation, and biodiversity, forestation could be very effective if combined with these other objectives.

Growing forests as a mitigation option competes with demand for land for agriculture and other uses. In a future, warmer climate, forests have been shown to potentially cover a larger fraction of the land (Leemans, 1992). The vegetation distribution model (BIOME; Prentice *et al.*, 1992) that is incorporated in the IMAGE 2 model shows a similar pattern for the period from the present (1990) to the middle of the next century (Table II). There is thus a larger potential for storing C in forested ecosystems in the future than at present. But this potential must be evaluated against the requirements for other land uses such as agriculture.

Table II lists the forest lands available now and those projected to be available in the future for different regions. The pattern is very clear: the only possible increases in the extent of forests are in the developed regions. In most other regions, large parts of the potentially forested areas will be used by agriculture, and little improvement from

TABLE II

Land suitability and availability (not used for other purposes) for the forestation mitigation option according to the IMAGE 2 simulation for the baseline scenario (data for current conditions derived from the BIOME model (Prentice *et al.*, 1992) and a land-cover database (Olson *et al.*, 1985))

Zone	Currently Suitable (10^6 km^2)	Currently Available (10^6 km^2)	Suitable in 2050 (10^6 km^2)	Available in 2050 (10^6 km^2)
North America	10.2	6.1	11.7	7.9
Latin America	11.1	6.2	14.3	6.5
Europe	3.9	1.3	4.2	1.4
Former Soviet Union	11.5	8.5	14.1	9.6
Africa	4.5	2.5	6.3	0.2
India	0.9	0.5	1.3	0.0
Southeast Asia	3.2	1.3	3.6	0.5
China	4.5	1.9	5.5	0.2

forestation will arise. On a smaller scale there will, of course, be many more opportunities to increase forest growth, such as planting trees in shelterbelts and agroforestry systems, that cannot be depicted in such a coarse-scale global model as IMAGE 2. The national emission inventories could assist in evaluating the potential of such efforts.

4. Conclusion

Several ecosystem processes that influence the flow of GHGs have been presented. These processes have very specific local and regional properties and respond differently to changes in land cover, climate, and atmospheric composition. However, these responses cannot be predicted from only local and regional conditions because global systemic properties of the Earth's system are involved. For example, CO_2 emissions from deforestation in one region can enhance the growth (and thus C sequestration) in other regions.

To evaluate future trends in atmospheric GHG concentrations, a proper modeling approach has to be developed, one that incorporates both the local characteristics of land and land use and the regional characteristics of society. This approach should be embedded in a global system involving the most important processes in the atmosphere and ocean. Several integrated assessment models (for example, IMAGE 2) have evolved (and are still evolving) along these lines. Such models can provide a comprehensive framework to address the regional to global issues of environmental change. Although these models are not suited to local and national assessments, local and national assessments could well be nested into the global approaches. The global approach, then, only defines the boundary conditions for and constrains the smaller scale assessments. For example, the global model provides the atmospheric CO_2 concentrations and climate patterns needed to define local growth rates of vegetation and thus determine the local C fluxes.

Data sets resulting from national inventories can assist in initializing and setting parameters for the different types of models, because the quality of the current underlying global data sets is poor (Leemans *et al.*, in press). These national inventories could thus well provide a good baseline data set; however, they do not allow (even if repeated regularly with a similar methodology) the evaluation of the effectiveness of mitigation strategies. This requires a systematic modeling exercise and the establishment of possible future baseline scenarios without mitigation options (Carter *et al.*, 1994), followed by analyses of implemented mitigation options. An integrated approach is required to address the sensitivity to feedback processes and compatibility of different options. The capacity to do such an analysis is currently being developed, and integrated assessment models could form the basis of such a comprehensive approach.

Acknowledgments

The invitation by the U.S. Country Studies Program on Climate Change to prepare this paper and to participate in the African Regional Workshop is highly appreciated. Stimulating discussion during the workshop and detailed comments by J. Fitzgerald, B. Braatz, and R. de Wijs greatly improved this manuscript. Research funds were granted to the IMAGE group by the Dutch Ministry of Housing, Planning, and Environment under contract MAP410 to the National Institute of Public Health and Environmental Protection (RIVM) and the National Research Programme "Global Air Pollution and Climate Change." The research presented here represents a contribution to IGBP-GCTE core research.

References

Alcamo, J. (ed.): 1994, *IMAGE 2.0: Integrated Modeling of Global Climate Change*, Kluwer Academic Publishers, Dordrecht, 318.

Alcamo, J., van den Born, G.J., Bouwman, A.F., de Haan, B., Klein Goldewijk, K., Klepper, O., Leemans, R., Olivier, J.A., de Vries, B., van der Woerd, H., and van den Wijngaard, R.: 1994, Modeling the global society-biosphere-climate system, Part 2: Computed scenarios, *Water, Air and Soil Pollution* **76**, 37–78.

Carter, T.R., Parry, M.L., Harasawa, H., and Nishioka, S.: 1994, *IPCC Technical Guidelines for Assessing Impacts of Climate Change*, IPCC Special Report, Geneva, 59.

Cramer, W., Leemans, R.: 1993, Assessing impacts of climate change on vegetation using climate classification systems, in Solomon, A.M., and Shugart H.H., (eds.), *Vegetation Dynamics Modelling and Global Change*, Chapman-Hall, New York, 190–217.

Edmonds, J., Wise, M., and MacCracken, C.: 1994, *Advanced Energy Technologies and Climate Change: An Analysis Using the Global Change Assessment Model (GCAM)*, Battelle PNL, Washington, DC, 19.

Jackson, R.B., Sala, O.E., Field, C.B., and Mooney, H.A.: 1994, CO_2 alters water use, carbon gain, and yield for the dominant species in a natural grassland, *Oecologia* **98**, 257–262.

Johansson, T.B., Kelly, H., Reddy, A.K.N., and Williams, R.H. (eds.): 1993, *Renewable Energy: Sources for Fuels and Electricity*, Island Press, Washington, DC, 1160.

Kirschbaum, M.U.F.: 1993, A modelling study of the effects of changes in atmospheric CO_2 concentration, temperature and atmospheric nitrogen input on soil organic carbon storage, *Tellus Series B—Chemical and Physical Meteorology* **45**, 321–334.

Leemans, R.: 1992, Modelling ecological and agricultural impacts of global change on a global scale, *Journal of Scientific and Industrial Research* **51**, 709–724.

Leemans, R., Cramer, W., and van Minnen, J.G.: 1995, Prediction of global biome distribution using bioclimatic equilibrium models, in Melillo, J.M., and Breymeyer, A. (eds.), *Effects of Global Change on Coniferous Forests and Grassland*, J. Wiley and Sons, New York (in press).

Leemans, R., Zuidema, G.: 1995, The importance of changing land use for global environmental change, *Trends in Ecology and Evolution* **10** 76–81.

Morita, T., Matsuoka, Y., Kainuma, M., Harasawa, H., and Kai, K.: 1993, *AIM: Asian-Pacific Integrated Model for Evaluating Policy Options To Reduce GHG Emissions and Global Warming Impacts*, draft report, Ibaraka, 26.

Olson, J., Watts, J.A., and Allison, L.J.: 1985, *Major World Ecosystem Complexes Ranked by Carbon in Live Vegetation: A Database*, Oak Ridge, Tennessee, 164.

Pepper, W., Leggett, J., Swart, R., Watson, J., Edmonds, J., and Mintzer, I.: 1992, *Emission Scenarios for the IPCC: An update. Assumptions, Methodology and Results*, Geneva, 115.

Prentice, I.C., Cramer, W., Harrison, S.P., Leemans, R., Monserud, R.A., and Solomon, A.M.: 1992, A global biome model based on plant physiology and dominance, soil properties and climate, *Journal of Biogeography* **19**, 117–134.

Rotmans, J., de Boois, H., and Swart, R.J.: 1990, An integrated model for the assessment of the greenhouse effect: The Dutch approach, *Climatic Change*, **16**, 331–356.

Schulze, E.D., and Mooney, H.A., (eds.): 1993, *Biodiversity and Ecosystem Function*, Springer-Verlag, Berlin, 525.

Sedjo, R.A. and Solomon, A.M.: 1989, Climate and forests, in Rosenberg, N.J., Easterling, W.E.-III, Crosson, P.R., and Darmstadter, J., (eds.), *Greenhouse Warming: Abatement and Adaptation*, Resources for the Future, Washington, DC, 105–119.

Smith, T.M., Leemans, R., and Shugart, H.H.: 1992, Sensitivity of terrestrial carbon storage to CO_2-induced climate change: Comparison of four scenarios based on general circulation models, *Climatic Change* **21**, 367–384.

Turner, B.L. II, Kasperson, R.E., Meyer, W.B., Dow, K.M., Golding, D., Kasperson, J.X., Mitchell, R.C., and Ratick, S.J.: 1990, Two types of global environmental change: Definitional and spatial-scale issues in their human dimensions, *Global Environmental Change* **1**, 14–22.

Turner, B.L., Moss, R.H., and Skole, D.L.: 1993, *Relating Land Use and Global Change: A Proposal for an IGBP-HDP Core Project*, IGBP Report, Stockholm, 65.

UNEP, OECD, IEA, and IPCC 1995, (United Nations Environment Programme, Organizations for Economic Cooperation and Development, International Energy Agency, Intergovernmental Panel on Climate Change), *IPCC Guidelines for National Greenhouse Gas Inventories*, Bracknell, 3 Volumes.

Van Campo, E., Guiot, J., and Peng, C.: 1993, A data-based reappraisal of the terrestrial carbon budget at the last glacial maximum, *Global and Planetary Change* **8**, 189–201.

Van Minnen, J.G., Klein Goldewijk, K., and Leemans, R.: 1995, Modelling the global carbon cycle in relation to climate change feedback processes and vegetation transitions, *Journal of Biogeography*, (in press).

Winjum, J.K., Dixon, R.K., and Schroeder, P.E.: 1993, Forest management and carbon storage—an analysis of 12 key forest nations, *Water, Air, and Soil Pollution* **70**, 239–257.

ASSESSING NATIONAL LIVESTOCK POPULATIONS FOR THE PRODUCTION OF METHANE EMISSION INVENTORIES

A. DE JODE

Appropriate Technology International, Global Livestock Producers Program (GLPP)
1828 L Street NW, Suite 1000, Washington D.C. 20036

Abstract. In an effort to accurately assess national methane (CH_4) production levels, many African countries are producing inventories of their national livestock populations. Conducting accurate livestock surveys is an expensive and time-consuming process, but is necessary because the most readily available data on livestock production are often incomplete or unsatisfactory. This paper looks at practical approaches to collecting data on livestock production and on the reproductive parameters for livestock, with special reference to African pastoralist populations, and makes qualified recommendations as to how inventory teams might be able to produce more accurate assessments of their total livestock populations. The methodology used during Nigeria's National Livestock Census (1990–92) serves as a basis for a discussion of the issues.

1. Introduction

Parties to the 1992 Framework Convention on Climate Change have agreed, under Article 4 of the convention, to develop, periodically update, publish, and make available to the Conference of the Parties national inventories of anthropogenic emissions of all greenhouse gases (GHG) not controlled by the Montreal Protocol. The emission of methane (CH_4) from digestive fermentation is one of the largest sources of global methane, while the production of methane in the digestive tract of ruminants is the greatest source of methane within the category of enteric fermentation. Once produced, CH_4 serves no purpose and is released into the atmosphere through eructation and defecation, as an end-product of the fermentation process, with a small amount of CH_4 entering the blood and being released through the lungs (Dougherty *et al.*, 1965).

This paper addresses the problem of defining African livestock populations in order to calculate their methane productivity levels. It concentrates particularly on methods of collecting productive and reproductive data on ruminants from the traditional livestock sector in a country. The pastoralist sector is highlighted in particular because of the relative magnitude of biomass in this sector compared with other livestock typically reared in an African country. Because many of the productive traits are similar across ruminant species, the focus of the data collection techniques described is on cattle, with the understanding that the techniques remain equally valid for other ruminant species. The author's experience in the execution of a national livestock census for the Federal Government of Nigeria is used as a basis for many of the examples of these data collection techniques. The advantages and disadvantages of the strategies available to inventory teams are discussed, and some recommendations proposed. The paper uses the four-part methodology proposed by the Intergovernmental Panel on Climate Change (IPCC) as its structure.

Environmental Monitoring and Assessment **38**: 217-229, 1995.

2. Methods

2.1. Information Required for Livestock Inventories

To assist countries in documenting their national GHG inventories, the IPCC has produced a manual that presents the methodology the IPCC views as suitable and acceptable for producing inventories. The *IPCC Guidelines for National Greenhouse Gas Inventories* (UNEP *et al.*, 1995) appears to have been adopted by the majority of inventory teams as the basis for their national livestock inventories. However, IPCC officials have stated that revisions to the IPCC guidelines over time are not only necessary but are actively encouraged.

The suggested methodology for estimating methane outputs from livestock in the current version of the IPCC manual can be summarized as follows:

1. *Documentation* of the total population for each livestock species
2. Division of each species into subgroups and the *characterization* of each of these groups
3. Estimation of the *emission factors for each subgroup* in terms of kilograms of methane per animal per year
4. *Multiplication* of the emission factors by the subgroup populations, then the summation across all the subgroups to estimate total emissions

Countries are not obliged to use the IPCC methodology and are encouraged to explore the proposed categorizations and emission factors. However, results and methodology must be "transparent" so the IPCC is able to adjust for any bias in inventories done by different countries. The first three stages in the suggested methodology all require a detailed knowledge of the national livestock population that may not be available to the inventory teams in all cases. The following sections examine each of these stages in turn, discussing the data sources that are likely to be available and strategies for overcoming data shortages.

2.2. Documenting Livestock Populations

2.2.1. Options available to country inventory teams

The two main methods for collecting accurate data on livestock populations are an annual *decentralized regional count*, undertaken periodically, and a *national count*, undertaken by one organization and in one single operation. The IPCC manual suggests that if no such data exist on which to base livestock population figures, inventory teams should use the appropriate Food and Agriculture Organization (FAO) yearbook. These yearbooks contain figures that are given to the FAO representative by the ministry responsible for livestock statistics in a country and tend to combine a variety of data sources, such as vaccination records, dipping records, slaughter records, head taxes, market records, and import/export records.

Annual regional counts are undertaken by several countries, including Ghana, which has a reputation for having refined an accurate and effective counting methodology. Countries that have undertaken full national counts are most likely to have done so for one particular species, usually cattle. Botswana, Niger, Nigeria, Chad, and Zimbabwe have all undertaken counts, although Nigeria is the only country that has undertaken a national inventory of *all* the livestock species and their respective breeds. A closer examination of the Nigerian National Livestock Census (NNLC) will show what is technically possible during a national census, although it must be recognized that undertaking such a census may be beyond both the scope and needs of many African states.

2.2.2. Case study of the Nigerian National Livestock Census, 1990–92

Nigeria's National Livestock Census was conducted as a prerequisite to the implementation of the Second Livestock Development Project, funded by the World Bank. The census was conducted by Resource, Inventory and Management (RIM) between 1990 and 1992. The objectives of the census were the following:

- Map and quantify the seasonal distribution and abundance of livestock species
- Characterize and assess the current status of Nigerian livestock breeds
- Collect representative herd productivity parameters
- Describe the overall distribution of production systems
- Identify major constraints to livestock production
- Highlight subjects and areas that would be appropriate for more detailed examination in subsequent phases of Nigeria's livestock development program
- Assess the size and composition of commercially managed livestock holdings

The methodology used in this census was developed by RIM over several years of work in Kenya, Chad, Niger, and Sudan. Nigeria was divided into 20 km transects on both longitudinal and latitudinal axes, forming 20 × 20 km blocks. This grid structure (2,280 grid cells in total) was used as the basis for an integrated sampling technique that employed both air and ground teams. Two planes flew systematic low-level reconnaissance flights (that is, planes flying at a constant speed and height above ground level) for an intense period during both the dry and wet seasons in an effort to examine the seasonal differences in livestock population. Observers counted actual numbers of livestock as well as other physical features of the livestock from a limited viewing panel, which was arranged on either side of the panel. The panel ensured an actual count of more than 5 percent of the surface of the country. The data collected determined for each grid cell the number and dominant color of all the visible livestock species (cattle, small ruminants, camels, donkeys, and horses), the number and type of all houses (pastoralist and nonpastoralist), the proportion of all vegetation types, and the occurrence of all water sources.

The groundwork was conducted by two sets of ground teams. Two village investigation teams spent a year traveling to each grid cell to determine the presence or absence of all types of livestock in sample villages, as well as the actual number of each species in a sample number of households. Additional information was obtained through interviews on draft power (including carting, oxen plowing, and water drawing), dairy activities and dairy products produced and sold, small-scale livestock fattening, and the purchase of mineral blocks and supplementary feeds for livestock. Statistically significant coefficients were then calculated to determine the number of livestock per dwelling for each grid cell; these were then extrapolated to the number of domestic rooftops that had been documented by the airteam.

Two further ground teams were charged with documenting *qualitative* livestock data—namely, productive and reproductive parameters—from the traditional pastoralist and agropastoralist sectors, which are the most significant livestock subsectors in terms of the country's total livestock biomass. The results of this national livestock census are now bound as four volumes containing a national synthesis, reports of the livestock resources from each state, livestock population estimates for twenty-four urban areas, and a national assessment of commercial livestock holdings (RIM, 1992). The usefulness of the census results for inventory teams charged with calculating methane emission is discussed in Section 3.

2.3. Categorization of Livestock Populations

Returning to the IPCC methodology, the second task of the inventory teams is to categorize the livestock populations. Table I shows the categorization that the IPCC guidelines suggest countries might use to group their national livestock population.

One of the problems with this categorization is that it does not take into account that in the context of African culture and economics the spectrum of livestock production varies considerably between the commercial (high-input) sector and the traditional, or pastoralist (extensive, low-input), sector. These separate production systems use different breeds and substantially different husbandry techniques, with the result that feed inputs, ruminant breeds, and mature body weights contrast sharply between the two. This is significant because the available research has shown that the main factors influencing the level of methane production from ruminants are the size, feed quality, and feed intake of the animal

TABLE I

Representative cattle categories and subcategories recommended by IPCC (UNEP *et al.*, 1995)

Main Categories	Subcategories
Mature Dairy Cattle	Dairy Cows used principally for commercial milk production
Mature Nondairy Cattle	Mature Females :
	Beef Cows: used principally for producing beef steers and heifers
	Multiple-Use Cows: used for milk production, draft power and other uses
	Mature Males :
	Breeding Bulls: used principally for breeding purposes
	Draft Bullocks: used principally for draft purposes
Young Cattle	Preweaned Calves
	Growing Heifers, Steers/Bullocks and Bulls
	Feedlot-Fed Steers and Heifers on High Grain Diets

(Blaxter and Clapperton, 1965; Crutzen *et al.*, 1986). The following sections and Figure 1 present a proposed alternative to the IPCC's categorization of national livestock populations.

2.3.1. Alternative categorization

Figure 1 shows the total livestock population divided into three main sectors: commercial, traditional (pastoralist), and peri-urban or village. Because in the majority of African countries the pastoralist sector is likely to be the mainstay of the livestock economy, gathering information on this sector is the crucial element of any inventory.

The commercial sector of many African states is less significant than the traditional sector in terms of the percentage population of the national herd and is, therefore, less relevant to this discussion. In addition, the commercial sector is far more amenable to data collection, because it is the subject of government records and because of the standardized nature of many of its livestock breeds.[1]

The third sector, village and peri-urban livestock, is also of less concern here, but it is worth pointing out that in many African countries this is the most rapidly expanding component of the livestock industry. Increasing urbanization has led to an expanding market for dairy products and meat, with urban consumption patterns becoming increasingly Westernized. The demand for "fast" foods and restaurant foods that contain a higher percentage of meat is rapidly increasing, encouraging entrepreneurs to set up operations that produce meat at a comparatively faster rate than the traditional sector is able to. Evidence of this growth is the fact that production and use of livestock concentrate

[1] Many commercial breeds are marketed at a certain age and at a certain weight. Liveweight gain for example is easily calculable, and the amount of energy that is required on a standardized diet of quantifiable quality can be calculated with reasonable accuracy.

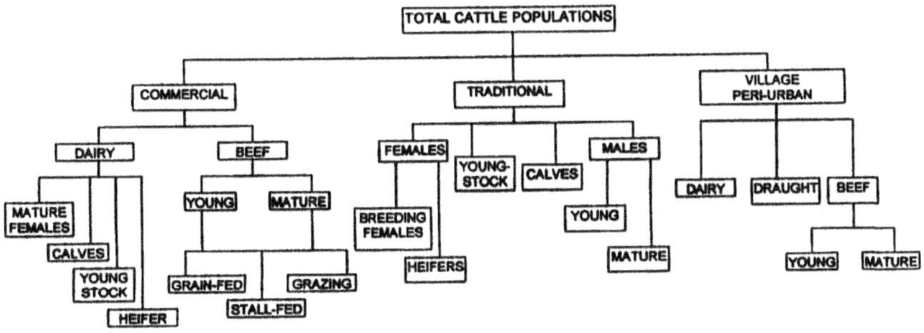

Figure 1. An alternative to characterization of cattle types.

feedstuffs are expanding in many African countries at a far faster rate than in Europe (FAO, 1984).

2.3.2. Characterization of pastoralist herds
Inventory teams seeking to characterize the pastoralist livestock population should first approach the livestock and veterinarian departments in the country's government. These departments will have varying amounts of information on the traditional livestock subsector in their country. There is also a huge body of literature on African pastoralists, some of which will undoubtedly be relevant to inventory teams. However, interviewing sample livestock owners in the field is still an essential part of the process, although the scale of the fieldwork will depend on the resources available to inventory teams.

Approaches to pastoralist communities. Collecting accurate livestock data obviously requires the full cooperation of the herdsmen, and it is sometimes very difficult to ensure that the information obtained is genuine, particularly if the majority of the cattle owners are nomadic or transhumant pastoralists who may be suspicious of people from outside their usual communities. It is extremely important that the initial contact establish a mutual trust, which can be achieved if the team employs mutually acceptable livestock extension workers, nongovernmental organizations, or other individuals who have a good working relationship with the pastoralists.

If possible, initial contacts should be made on neutral ground, rather than at the pastoralist encampment itself, so that the first meeting can be spent making the introductions that are necessary to ensure the full cooperation of the cattle owner. It is important to communicate exactly what information is required, how any information is to be recorded, and what the information will be used for, answering all queries honestly and ensuring anonymity if it is requested. Experience in Nigeria showed that some cattle owners change their minds several times about their willingness to participate in interviews and, on many occasions when they do agree to an interview, become bored with the monotony of the questions. Interview times at the encampment should take into account milking patterns, grazing times, and so forth, and should not exceed 1.5 hours.

Assessing the herd structure. Herd and flock demography are difficult to chronicle. Pastoralists, in particular, are unwilling to let government workers count their animals directly, fearing, often justifiably, that the information will be used to impose a livestock head tax. In Nigeria, the inventory team managed to obtain a reasonable picture of the herd and flock demography without direct questioning by using the following technique:

1. Cattle owners were asked about the number of mature females in the herd, rather than actual herd numbers. If this information was not forthcoming, the number of breeding females in the herd was counted inconspicuously over the course of the interview. (Breeding females are here defined as mature female cows that have calved at least once.)
2. Individual breeding females were then randomly selected from the herd, and a breeding females case history form was used to document details of their personal history as well as the history of their progeny. The date of birth, sex, and presence or absence of their latest calves were recorded, along with reasons for the exit of the progeny from

the herd, for example, death, sale, transfer, or gift. The interviewee was then asked to identify for each cow the calf born *before* the most recent calf, and *its* date of birth, sex, and fate were recorded. In this way, the ages of all the progeny and the exit ages of all animals in the herd were recorded, and the herd structure accurately plotted.

3. Many pastoralists take a longer term view of events and tend to judge animals according to state of maturity, physiological state (that is, weight or height), reproductive history, and barter value rather than actual birth dates or other statistics (Iles, 1994). To increase the accuracy of birth dates, the owners were asked to relate these to a religious festival or seasonal occurrence (agricultural or climatic).

By compiling a number of these breeding-female histories from an area with similar livestock husbandry systems or from herds with the same breed, it is possible to gather a considerable amount of productive data on the regional herd and to build up a picture of the national herd demography. Such a technique could be used by the IPCC inventory teams, with questions adapted to their precise needs. Sample herd data, divided into separate age categories, can then be correlated with age and weight data collated from secondary sources for livestock in other regions.

2.4. Determining Emission Factors per Subgroup

The third stage of the IPCC methodology requires that inventory teams estimate the emission factors for each of the subgroups into which the population has been categorized. This estimate requires data on average daily feed intake. Although relatively easy to obtain from the commercial sector, these data are often unavailable in the traditional sector and, therefore, must be estimated through the calculation of weight, weight gain per day, energy expenditure through feeding, milk production, work performed, and feed digestibility (IPCC/OECD, 1994, p. 4.6). The following are techniques for calculating these from a sample population:

Animal weight. In the absence of a weighbridge, inventory teams can calculate cattle weights using a formula that exploits the relationship between a series of cattle body measurements and the animal's liveweight (LW). The measurements most commonly used for this formula are body girth, body height, and body length. An appropriate formula may have already been calculated for a particular breed during livestock production research or comparative husbandry studies. If so, this would be the logical starting point for the inventory teams.

Collecting livestock body measurements can disrupt the herd, which in turn will upset the herd owner. Therefore, measurements should always be taken at the end of any visit. Enlisting the help of people who are in day-to-day contact with the herd (women from the family, the regular milker, or a child) is also encouraged. Using a tape measure (preferably cloth), body girth is measured by passing the tape around the animal's body and collecting it just behind the shoulder blade. Body height is measured by letting the tape touch the ground and, with a book or stick resting horizontally on the top of the shoulder blades, reading the height. Body length is measured from the midpoint of the shoulder blades on the spine to the midpoint of the pin bones along the spine.

Daily liveweight gain (DLWG). Inventory teams can estimate the average weight gain per day (exclusive of seasonal bodyweight fluctuations) from a large sample of animals by taking the weights of animals of different ages, extracting their birth weight, and dividing this by their ages (in days). DLWG for different herd categories (calves, youngstock, yearlings, heifers, and so forth) can also be estimated by subtracting the weight of a representative animal in a younger category, from the weight of an animal in an older category. The average liveweight gain per day gives an indication of how the animal's feed has deviated from the optimal diet. Because the animal is physically limited by the amount of dry matter it can consume in a day, this liveweight gain is also an indication of the quantity and quality of the feed intake.

Energy expenditure through feeding. Animals reared in the open expend more energy to obtain food than do stall-fed animals. IPCC states that animals grazing over very large areas (epitomized by the African pastoralist herds) require an additional 37 percent of the net energy[2] they require for maintenance, and that this should be taken into consideration in the calculations. (Maintenance is the apparent feed energy required to keep the animal in energy equilibrium; that is, there is no gain or loss of energy in the body tissues.) Maintenance is the apparent feed energy required to keep the animal in energy equilibrium; that is, there is no gain or loss of the body tissues.) Many pastoralist herds walk enormous distances, particularly during the dry season when their body weights are often lowest. As a result, the majority of these animals are in a negative energy balance during this period, and it would be useful to explore the differences in exertion, and subsequent energy requirements, between an animal in a negative energy balance and an animal in a maintenance or plus-maintenance status.

Diet Quality. IPCC has given feed digestibility guidelines for different feed sources: grazing rangelands, grazing good pastures, crop residues, concentrates, and so forth. These digestibility levels are used to calculate methane conversion rates, which, along with estimates of feed intake, become the emission conversion factors for each livestock sector. However, if the seasonal body-weight fluctuations, which may be as much as 20 percent (Barrett and Larkin, 1974), can be linked to the feed quality at the time, this information, coupled with the dry matter intake limitations, will give a far more accurate picture of the net energy intake of these animals.

It may be possible to obtain a more accurate estimate of the emission conversion factors in the traditional sector by investigating the annual feeding regime of traditionally managed herds, which will vary little among regions. Block periods of distinct consumption (and the respective qualities of feed) can be directly correlated to the seasonal weight gain and loss of herd or flock animals. Estimating the feed intake of mature animals without considering the inefficiencies of the energy expenditure in annual liveweight gain and loss grossly underestimates the situation and, therefore, underestimates the sector's methane gas emissions.

---

[2]Net Energy = Gross Energy – Fecal Losses – Urinary and Combustible Gas Losses – Heat Increment.

3. Results and discussion

3.1. Livestock Population Data

The main objective of inventory teams is to obtain national livestock population data that is as accurate as possible. It is widely recognized that the FAO sources of livestock data fail to take into account a number of factors. First, a high percentage of the livestock population does not come into contact with veterinary services (particularly in the pastoralist community), and there will be no data at all on certain groups of livestock, such as small ruminants. Second, a large amount of slaughtering is conducted "informally" in many African countries. Third, these records fail to appreciate how fluid national boundaries are for livestock populations, particularly for large ruminant populations exploiting a wide area of natural grazing. It is therefore essential that more accurate sources of data are used.

The most effective means of obtaining the required data is to conduct a national census. However the accurate documentation of a country's total livestock population is not only technically and logistically difficult but may also be politically controversial, particularly where a district or state's livestock operational/development budget is based on its percentage of the national livestock population. In Nigeria, the livestock census was conducted by a single outside organization, ensuring that the data collected were independent of any bias.

Because the Nigerian census was based on the use of individual grid cells, a large amount of data could be linked to a geographical location, rather than to any artificial boundary. The density of livestock in each of the 2280 individual grid cells was documented, and through the integration of the air and ground data, these livestock densities could be defined by breed as well as by agroecological zone. The fact that the survey was undertaken during both the wet and the dry seasons meant that the density counts could compare the livestock distribution on a seasonal basis.

National livestock counts have the key advantage of ensuring that a concentrated, intense "blanket" operation is undertaken, thus eliminating the risk of double counting. This is particularly important in countries where there is a significant pastoralist population moving between regions in search of grazing. The Nigerian census, however, was a very detailed operation that covered every aspect of domestic livestock rearing in a large and diverse country. This scale of investigation is not essential for calculating CH_4 emissions and could not be expected from inventory teams, nor made possible through funding provided by the Intergovernmental Panel on Climate Change/United Nations Environment Programme (IPCC/UNEP). However, it may be feasible for the costs of a more simplified census to be shared by a country's national livestock department and another ministry that might be in need of aerial data—for example, wildlife departments. Census data are invaluable for justifying livestock-based development projects and are important for monitoring impacts of drought, and so forth. The high costs of acquiring these data, however, mean that in reality a national livestock census may only be feasible in countries whose economies depend heavily on livestock.

Inventory teams following the IPCC guidelines are unlikely to rely entirely on FAO-type statistics, but neither will they have immediate funds for a national census. The majority will have access to disparate datasets on ruminant livestock populations and productivity, collected through either regional livestock surveys or project-specific research. Regional counts are cheaper and are, therefore, likely to have been conducted more frequently than national surveys. They also have the advantage of being able to provide a focused, regional dimension to an assessment of the livestock population. Their main disadvantage is that regional data varies in quality, so that extrapolation from regional data to a national picture may require additional knowledge that may not be immediately available to inventory teams.

To acquire data that are useful for the national inventory and, hence, for emissions calculations, it is highly likely that the inventory teams will need to conduct further fieldwork. This in turn depends on the funds available for conducting detailed investigations of sample cattle categories, the team's knowledge of the subject, and the amount of time available for investigating the research literature to fill in the gaps in the team's knowledge. It is important that teams collaborate with national and international institutions that are able to help them find all the relevant information they will need—for example, through on-line databases.

3.2. Characterization of Livestock Subgroups

Characterization of subgroups is necessary to calculate a national subsector biomass. The categories and subcategories of cattle types proposed in the IPCC guidelines were not designed specifically for use in the context of African culture and economics, and inventory teams may do well to consider developing their own classification systems. The fact that a high percentage of the national herd is likely to be reared under an extensive pastoralist system and will, therefore, have different feed inputs and methane emissions than commercial (intensive) livestock is a crucial consideration. Small-scale, backyard, peri-urban livestock should also be considered in the development of categories. Livestock departments, in particular, should take note of this sector while it is still relatively small, to make an objective assessment of its temporal expansion and importance to the peri-urban economy.

Further defining the national herd through fieldwork is important for verifying proposed classifications, but a prerequisite for any fieldwork is a thorough investigation of all the relevant literature available, including literature from neighboring countries in which the same breeds are found. Gathering data in the field is the most expensive part of any data collection exercise, and if this has already been done by research institutions or others, costs can be reduced considerably.

Fieldwork among pastoralists is not only time-consuming and physically demanding, but the quality of the data itself is only as good as the mental recall of the herdsman being interviewed. During the Nigerian census, a number of the case histories had to be discarded because the dates given were not feasible. Good data on herd structures can be obtained, however, and can be seen for one of the Nigerian cattle breeds in Figure 2, where the

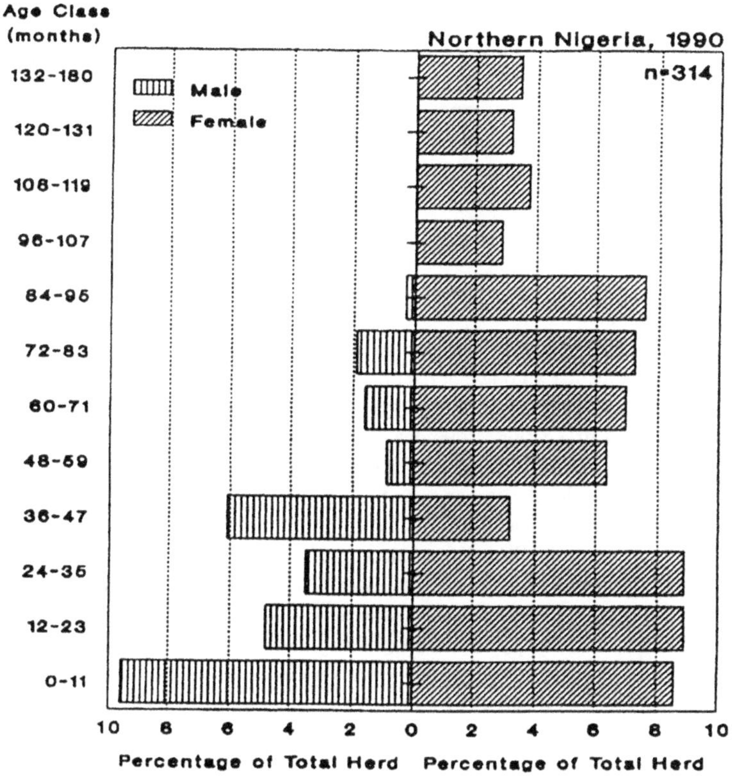

Figure 2. Herd demography of Bunaji cattle owned by transhumant pastoralists in Nigeria, 1991.

herd demography is clearly defined into distinct age categories. This diagram shows clearly at which ages the different sexes either stay or leave the herd for different reasons. These data could also have been estimated from secondary data that documented breed-specific age-to-weight ratios.

3.3. Emissions Estimates for Subgroups

The individual breeding-female history forms used to obtain the data in Figure 2 can also be used to obtain a great deal of additional data pertinent to estimating average daily feed intake in order to calculate emission factors. Structured interviews also increase understanding of the wider constraints to husbandry practices and livestock production, providing insight into herd off-take priorities (including the fate of off-takes), livestock disease (morbidity, symptoms, treatment, and outcomes), feedstuff preferences, grazing pressures and disputes, and historical changes in the herd. The breeds of cattle kept may often be an indication of the ethnic group of the pastoralist owners, information that is important to teams constructing a national picture of the herd. Combining different methods of livestock data collection will generate an accurate baseline of productivity data that can show variations in productivity over time. The technique is also useful for assessing the effectiveness of any future intervention, be it a feed or mineral supplement, medication, or other.

4. Conclusions

The IPCC guidelines propose that signatories of the Framework Convention on Climate Change gather together all relevant data on their livestock populations and the categories into which they can be divided. The ideal means of determining the scale and documenting the characteristics of a nation's livestock sector is likely to be a national livestock census, although the high cost of conducting such a census means this method is truly feasible only for those countries that depend heavily on livestock. The alternative is to collate all the available research data and to complete the gaps by designing specific, well-structured field investigations that allow diverse data to be collected at lower cost, particularly if this cost can be shared with other departments or if help can be enlisted from national or international nongovernmental organizations. By conducting or commissioning well-targeted fieldwork, inventory teams charged with defining methane emissions can reduce their dependence on disparate data sets from various sources.

Until apparatus that is appropriate for measuring methane production in a tropical field situation is more widely available, inventory teams must continue to use mathematical formulae to calculate actual methane emissions. The use of formulae-generated estimates remains the only means available. As the U.S. Country Studies Program and inventory assessors refine the inventory process, it is hoped that funds may be made available for specific field studies to clarify, verify, or refine datasets that already exist, and for research studies to reexamine the accuracy of the estimates and formulae that have been used in the current methodology (Dixon *et al.*, 1996).

Increasing both the amount and accuracy of the livestock data from all the signatories of the Framework Convention is important not only for the immediate purposes of estimating methane emissions, but also for the longer term objective of finding the means by which these emissions can be mitigated.

References

Barret, M.A. and Larkin, P.J.: 1974, *Milk and Beef Production in the Tropics*. Oxford, U.K.: Oxford University Press, 169.

Blaxter, K.L. and Clapperton, J.L.: 1965, Prediction of the Amount of Methane Produced by Ruminants, *British Journal of Nutrition* **2419**, 511–522.

Crutzen, P.J., Aselmann, I., and Seiler, W.: 1986, Methane Production by Domestic Animals, Wild Ruminants, Other Herbivorous Fauna, and Humans, *Tellus* **38B**, 271–284.

Dixon, R.K., Sathaye, J.A., Meyers, S.P., Marera, O.R., Makosov, A.A., Toure, S., Makundi, W., and Wiel, S.: In press. Greenhouse gas mitigation strategies: preliminary results from the U.S. Country Studies Program, Ambio.

Dougherty, R.W., Mullenaux, C.M., and Allison, M.J.: 1965, *Physiology of Digestion in the Ruminant*, Dougherty, R.W. (ed.), London: Butterworth, 159.

Iles, K.: 1994, The Progeny History Data Collection Technique: A Case Study from Samburu District, Kenya, *RRA Notes, Number 20, Special Issue on Livestock*. IIED.

FAO (Food and Agriculture Organisation): 1984, Livestock Development in Developing Countries and Implications for Consumption and Trade of Feeds, *World Agricultural Economics & Rural Sociology Abstracts*, 026–00598.

Resource, Inventory, and Management Ltd.: 1992, *Nigerian Livestock Resources*. Four volume report to the Federal Government of Nigeria.

UNEP, OECD, IEA, IPCC (United Nations Environment Programme, Organisation for Economic Co-operation and Development, International Energy Agency, Intergovernmental Panel on Climate Change): 1995, *IPCC Guidelines for National Greenhouse Gas Inventories*, IPCC, Bracknell, 3 Volumes.

CARBON STORAGE IN EUCALYPTUS AND PINE PLANTATIONS IN SOUTH AFRICA

S.I. CHRISTIE[1] and R.J. SCHOLES[2]

[1] *Division of Forest Science and Technology, CSIR, Private Bag X11227, Nelspruit 1200, South Africa.* [2] *Division of Forest Science and Technology, CSIR, P.O. Box 395, Pretoria 0001, South Africa.*

Abstract. Carbon (C) is stored by plantation forests either when ecosystems with a low C density (such as tropical grasslands) are afforested or when timber is converted to semipermanent products. If the afforestation rate is relatively constant and the plantations are not harvested immediately upon reaching maturity, the amount of C stored in trees as a result of afforestation can be calculated by a simple "static" approximation. Rotation forestry requires a mean C storage method that averages C density over the rotation. Plantation forestry as practiced in South Africa requires a more detailed dynamic approach that accounts for time-varying rates of afforestation and the age-dependence of C accumulation rates in plantations. To determine C storage in products, the output of long-lived plantation products and their C content once all processing losses are accounted for must be known. The South African case study shows that new afforestation stored approximately 2.54 Tg C in 1990, and storage in forest products accounted for an additional 1.15 Tg C. Together, these two activities offset approximately 3.8% of the carbon dioxide emissions from South Africa.

1. Introduction

Carbon dioxide (CO_2) that is removed from the atmosphere and stored as carbon (C) may be considered to be sequestered in terms of the United Nations Framework Convention on Climate Change, and can be offset against the CO_2 emissions for that country. One mechanism of C storage is through its assimilation in vegetation. Numerous authors have estimated the potential of forests and agroforestry management practices to store C (Dixon *et al.*, 1994; Harmon *et al.*, 1990; Schroeder, 1992; and Winjum *et al.*, 1992).

When an ecosystem with a low C density (the amount of organic C per unit of land area, in Mg C ha^{-1}) is converted to an ecosystem with a high C density, there is a net increase in C storage equal to the difference in C densities over a period equal to the time taken to reach the new state. Such conversions take place in southern Africa when grasslands or sparse savanna woodlands (low-C-density ecosystems) are afforested to produce pulpwood, mining timber, sawtimber, or fuelwood. The amount of C stored will depend on both the forest management practices (that is, species planted, espacement, thinning, and rotation length) and the bioclimatic region (Winjum *et al.*, 1992).

Plantations are a common land use in Africa, but constitute only a small fraction of the land surface (1.2% in South Africa, one of the most extensively planted African countries). They almost invariably consist of exotic (that is, non-African) species of fast-growing trees, generally drawn from the genera *Pinus* or *Eucalyptus*. Most of these plantations are managed on a rotational basis; when the growth rate of the trees begins to decline, they

Environmental Monitoring and Assessment **38**: 231-241, 1995.

are clearfelled and another crop of trees is planted. Because of the favorable growing conditions in the tropics, the rotation length can range from 6 to 25 years, depending on the nature of the end product.

Even though the plantation is eventually felled, it nevertheless constitutes a net C storage if the mean C density over the rotation is greater than that of the vegetation it replaces. Furthermore, if any tree products are processed into a form where C is stored for a lengthy duration (for C-containing products of the fossil fuel industry, the Intergovernmental Panel on Climate Change considers a minimum life of 20 years), they constitute storage whether or not the forest area is expanding (or even if it is contracting, because in that case they must be subtracted from the C emission due to land use change).

The objective of this paper is to demonstrate, through the example of South Africa, the methods used to calculate C storage from the activities of the plantation forestry industry.

2. Materials and methods

2.1. Carbon Storage due to Afforestation

In the absence of more locally specific data, the C density of plantations can be based on the published averages (for example, OECD, 1991; IPCC/OECD, 1994; UNEP *et al.*, 1995). Most countries in which forestry is practiced will have data on final harvestable roundwood yields, which can be used to improve the default estimates. The total biomass of forests is about 50% greater than the mass of extractable timber at harvest because total biomass also includes branches, bark, leaves, and roots. Many tree species have been used in planation forests in South Africa. The softwood plantations are predominantly *Pinus patula*, while the hardwoods are predominantly *Eucalyptus grandis*. These two species have been taken as representative of all softwoods and hardwoods in South Africa, respectively.

The C density of the biomass standing crop can be calculated as

$$C_i = V_{stem}\rho_{wood}F_{carbon}F_{stem}^{-1},$$

where

C = tree biomass C density (Mg C ha^{-1}),
V_{stem} = stem wood volume (m^3 ha^{-1}),
ρ_{wood} = density of wood (the density of wet wood is 0.86 Mg m^{-3} for *E. grandis* and 0.88 Mg m^{-3} for *P. patula;* the moisture content of wood at harvest is 66% for *E. grandis* and 115% for *P. patula* (van Vuuren *et al.*, 1978)),
F_{carbon} = fraction of oven-dry mass that is C (assume 0.5 for all species), and
F_{stem} = fraction that stemwood contributes to whole tree biomass ha^{-1} (0.667 for *E. grandis* and 0.704 for *P. patula;* see Table I).

Values for the C contents of plant tissues vary in the literature from 45% to 54% (Kinerson *et al.*, 1977). A conversion factor of 50% is commonly used to estimate the organic C content of plant tissues (Brown and Lugo, 1982; Harmon *et al.*, 1990). Table I

TABLE I

Biomass components of *P. patula* and *E. grandis* in southern Africa at harvest maturity (Christie, 1991; Herbert and Robertson, 1991; Morris, 1992)

	Percent of total dry biomass		
	Pinus patula	*Eucalyptus grandis*	
Component	20 years	8 years	7 years
Bole wood	70.7	66.6	75.2
Branch wood	10.1	9.4	2.7
Bark	5.9	7.5	7.1
Leaves	1.5	3.6	2.3
Roots	11.5	(13.0)[a]	(12.7)[a]

[a]Root biomass was not measured but is estimated to constitute 15% of the total aboveground biomass.

presents the data used to estimate F_{stem} for typical *Pinus* and *Eucalyptus* species growing in plantations in South Africa.

In South Africa, afforestation replaces montane grasslands, which have an aboveground biomass averaging 6 Mg DM ha^{-1} (Everson *et al.*, 1988). Assuming an equal amount of biomass belowground and 50% C content, this converts to a C density of 6 Mg C ha^{-1}. The method as applied assumes that soil C is not affected by the land-use change. This is certainly not true, but there are insufficient data to make a generalization about the magnitude (or even the sign) of the changes.

Three methods are described below for calculating C storage by plantations. The first two are static methods that do not account for the history of afforestation, while the third (dynamic) method does.

2.1.1. The static method using final carbon density

This is the simplest approach, and is the default in the *IPCC Guidelines for National Greenhouse Gas Inventories* (UNEP *et al.*, 1995). In essence, it credits the country with the full final biomass of the plantation on the date on which it is planted (about a decade or two before it actually accrues). This is not a serious source of error if the area planted annually is relatively constant and the trees are not harvested until some long time after they reach maturity. The calculation is as follows:

$$C_s = \sum_{j=1}^{m} A_j (C_{p,j} - C_g),$$

where

C_s = total C stored (Mg),
A_j = area planted per year of plantation type j (ha),
C_p = C density at maturity of the plantation type j (Mg C ha^{-1}),
C_g = C density of the preceding land use (Mg C ha^{-1}), and
m = number of different plantation types (including different species and management regimes).

Table II presents the estimated C density at maturity for *P. patula* and *E. grandis* sawtimber and pulpwood management regimes for both coniferous and broadleafed forests. The area planted per year is obtained from national or regional statistics (Figure 1). Where it is highly variable, a 3- or 5-year average centered around the reference year should be used.

2.1.2. The static method using mean carbon storage

If the trees are harvested before or soon after they reach maturity, the mean carbon storage (MCS) method (Schroeder, 1992) is more appropriate than the peak biomass method described above. The MCS method determines the average amount of C on a site over a full rotation. It assumes that the plantation management system is sustainable and no yield reduction in the second and subsequent rotations. The calculation is made by summing

TABLE II

Carbon density at maturity in average South African plantations of *P. patula* and *E. grandis* under different management regimes (van Vuuren *et al.*, 1978)

Plantation type	Rotation length (years)	Final yield (m^{3}ha^{-1})	Mean Annual Increment (m^{3}ha^{-1}yr)	Final C density (Mg C ha^{-1})	Mean C storage (Mg C ha^{-1})
African coniferous forests (OECD 1991)[a]36.8–67.6					
P. patula, sawtimber	25	217 (172–563)	14.8	62.6	32.6
P. patula, pulp	15	206 (166–439)	14.8	59.2	31.7
African broadleafed forests (OECD, 1991)[a]72.8–117.0					
E. grandis, sawtimber	25	695 (439–695)	31.6	269.9	140.3
E. grandis, pulp	10	122 (114–465)	21.3	47.4	26.1

[a]Aboveground biomass only.

Note: The Mean Average Increment (MAI) is determined at age 20 for sawtimber regimes and at rotation length for pulpwood. The MAI values used are the area-weighted means for South Africa (FOA, 1994). The final yields were from tables in Loveday and Kassier (1993); and the range is given in brackets.

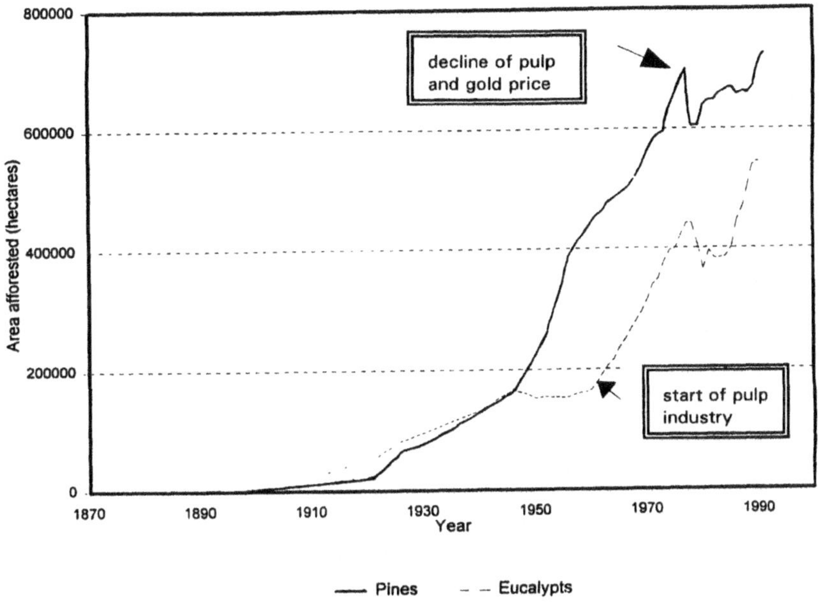

Figure 1. Cumulative afforested area in South Africa.
Sudden changes in the rates of afforestation can usually be ascribed to economic factors.

the C content of the standing crop for each year in the rotation and dividing by the rotation length.

$$C_m = n^{-1}\sum_{i=1}^{n} C_i ,$$

where

C_m = mean C stored (Mg C ha^{-1}),

C_i = C density of the standing crop for year i (the values are obtained from equations 6 and 7), and

n = rotation length (years).

Table II also presents the C densities for *E. grandis* and *P. patula* calculated using the MCS method. Multiplying the mean C density by the mean rate of afforestation for the respective species and management regimes is an improved static method estimate of C stored, because it allows for the fact that the plantations are harvested rotationally.

$$C_s = \sum_{j=1}^{m} C_{m,j} A_m ,$$

where

C_s = total C stored (Mg),
$C_{m,j}$ = mean C density for plantation type j (Mg C ha^{-1}), and
A_m = mean rate of afforestation over a reference period.

2.1.3. The dynamic method

If the rate of afforestation varies greatly from year to year, the carbon density increment in a given reference year due to growth in that year must be tracked by each age-cohort of trees of each plantation type. This is known as a *dynamic* method.

$$C_s = \left(\sum_{i=1,j=1}^{m,n} C_{i,j} A_{(r-i,j)} \right) - C_g A_r$$

where

$C_{s,r}$ = C stored in reference year r (Mg C),
$C_{i,j}$ = C density of plantations of age i and type j (Mg C ha^{-1}), and
$A_{r-i,j}$ = Area of type j planted in year $r - i$ (ha).

The growth rate of a given cohort for a given year can be calculated using an empirical growth function that starts at close to zero on the date of planting and reaches a given final biomass at a specified harvest date. Examples of such functions, determined from forestry trial data in South Africa, are shown below.

Harrison *et al.* (1994) give the following result for *P. patula* in Nelshoogte, South Africa:

$$V_{stem} = V_{final}\left(1 - \exp\left(-0.1411\left(1 - 0.6782\exp(-0.0013TP)\right)i\right)\right)^{3.0935}$$

with $n = 218$, $R^2 = 0.99$, and $S_{y,x} = 44.72$ m^3.

Kotze (1991) gives the following result for *E. grandis* in Langepan, South Africa:

$$V_{stem} = V_{final}\left(1 - \exp\left(-(0.063471 + 0.0000089299TP)i\right)\right)^{1.85873}$$

with $n = 86$, $R^2 = 0.94$, and $S_{y,x} = 52.65$ m^3.

For these equations,

V_{stem} = Utilizable timber volume (m^3 ha^{-1}), and
V_{final} = Final stem volume (m^3 ha^{-1}) under optimal conditions (657.94 for *P. patula* and 274.79 + 41.7375 × S for *E. grandis*) in South Africa, where *S* is the site index defined as the height reached after 10 years growth in pulpwood (20 m) or 20 years in saw wood (30 m).
j = Stand age (years).
TP = Surviving stems per hectare (for *P. patula* sawtimber is 1200 at planting, thinned to 650 at year 9, 400 at year 14, and 250 at year 19; for *E.*

grandis sawtimber is 1,200 at planting, thinned to 750 at year 4, 500 at year 7, and 300 at year 10. Pulpwood has no thinnings and commences at 1,200).

Although these functions are in volume units, they can be expressed in C density units by substituting for volume in Equation 1 (Figure 2).

2.2. Carbon Storage in Timber Products

Carbon storage in timber products is estimated by summing the C content of final forestry products have an expected life of more than 20 years. In South Africa most construction timber falls into this category, as well as timber destined to be permanently buried underground as mine supports. The C stored in timber products can be calculated as:

$$C_p = \sum_{k=1} V_k \rho_k \in_k F_{carbon} ,$$

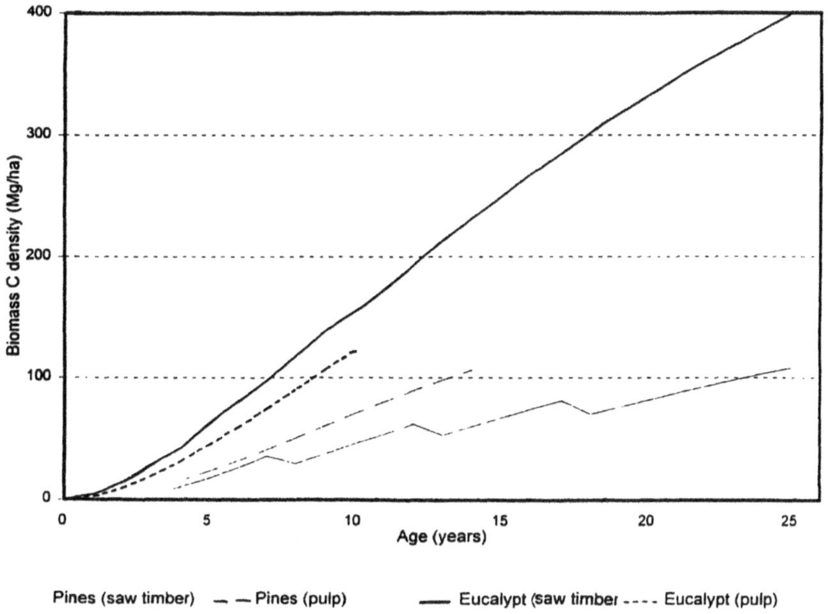

Figure 2. The biomass C density in the four main plantation types in South Africa. The dips in C density in the sawtimber plantations are due to thinning operations.

where

C_p = C stored in timber products (Mg yr^{-1}),
V_k = annual volume of roundwood timber used for product k (m^3 yr^{-1}),
ρ_k = density of air-dried timber product k (Mg m^{-3}), and
$\mathring{a}_k$ = milling efficiency (fraction of roundwood that ends up in product k).

Estimates of the amount of C stored in sawtimber, board and panel products, and mining timber are given in Table III Volumes are obtained from LHA Management Consultants (LHA, 1993). If timber output is quoted on a volume basis, it must be converted to an oven-dry mass basis. The mean density at 10% moisture content (air dry) for *P. patula* is 0.45 Mg m^{-3} (range is 0.35 – 0.61); for *E. grandis* it is 0.57 Mg m^{-3} (range is 0.5 – 0.8). If the product data are quoted as roundwood (that is, unsawn logs), they must be reduced according to the milling efficiency (the ratio of the mass of sawn timber to input roundwood is 46 to 55% for pine and 26 to 30% for eucalyptus). The milling efficiency for mining timber is estimated at 80%. In general, wood-based products have a C content of 50% by mass. It is assumed that sawmill residue is generally burnt, and therefore C is not stored. However, this is not strictly true, as increasingly large amounts of waste from sawmills are utilized in the board and panel manufacturing sector.

3. Results and discussion

The time series of afforestation in South Africa is illustrated in Figure 1. Wattle (principally *Acacia mearnsii*) was formerly extensively planted but now comprises less than 10% of the afforested area and 6% of new plantings. The impact of economic conditions on the rate of afforestation is clear. The increase in planted area is likely to continue, but ultimately will be limited by environmental and water-use considerations.

Table IV summarizes the forestry C storage calculations for South Africa using different methods. The simplest static approach, which uses the peak biomass to calculate the C

TABLE III

Carbon storage in long-lived (20-plus years) timber products

Product	Volume of roundwood used (× 10^6 m^3)	Oven-dry density (Mg m^3)	Moisture content of roundwood (% of air dry mass)	Milling efficiency (%)	C stored (Tg C y^{-1})
P. patula, sawn timber	3.80	0.45	115	50	0.39
E. grandis, sawn timber	0.05	0.57	66	28	0.05
Boards and panels	0.13	—	—	100	0.13
Mine supports	2.60	0.57	80	0.58	
Total	1.15				

TABLE IV

Carbon storage in 1990 in *P. patula* and *E. grandis* plantations (Tg C y^{-1}) growing under sawtimber and pulpwood regimes, calculated using different methods

Plantation type	Static methods: Peak C density	Static methods: Mean C storage Tg C	Dynamic method
P. patula, sawtimber	0.14	0.11	0.08
P. patula, pulpwood	0.55	0.44	0.14
E. grandis, sawtimber	1.16	0.59	0.85
E. grandis, pulpwood	0.87	0.44	1.47
Total	2.72	1.58	2.54

Note: For static methods, the area of new afforestation was 11 420 ha of pines and 22 700 ha of eucalyptus. It was assumed that 19% of both was destined for sawwood and 81% for pulp.

density, overestimates C storage in rotational plantations by a factor of nearly two relative to the static method using MCS (Winjum *et al.*, 1992).

The dynamic method is most appropriate for situations (such as in South Africa) where the rate of afforestation varies over time. The static MCS method underestimates C storage in South Africa by 37%, because the current rate of afforestation is less than it was during the last decade. The similarity of the estimates by the static peak C density method and the dynamic method is purely coincidental.

The calculations outlined above do not allow for changes in C density other than changes due to an increase in biomass. It is more than likely that the C stored in soil organic matter also changes. Soil C is often assumed to be higher under forests than under grasslands, but this is not necessarily true; grasses translocate a much larger fraction of their C belowground than forests do. On the other hand, the amount of C stored on the surface of the soil as litter can be large in forests, especially under pine plantations. In the Eastern Transvaal of South Africa, the litter accumulation can reach 150 Mg DM ha^{-1}, or about half of the stem biomass. However, until there are more reliable data on the magnitude of soil and litter C pools in tropical plantations, it is conservative to ignore them in the calculations.

The C stored in various long-lived timber products is given in Table III. The amount of C stored in these products is comparable to that stored in biomass. The default assumption in the *IPCC Guidelines for National Greenhouse Gas Inventories* (UNEP *et al.*, 1995) is that timber products are merely replacing existing timber products that are decaying or burned, and therefore should not be considered a net storage. However, in rapidly growing and urbanizing countries such as South Africa, the national pool of timber is increasing. Most sawn timber and board products are used for construction in South Africa, so it is reasonable to expect them to have a long life. The mining support timber is left underground when the mines are abandoned, and can therefore be considered to be securely stored C.

It is unclear how to consider paper pulp for emission inventory purposes, because it is a partially stored product. Although most paper products have a relatively short life (less than 5 years), a large fraction of these disposable products is recycled in some countries (31% in South Africa in 1992 (Gerischer, 1993). Much of the paper that is not recycled ends up in landfills, where it may have a long life. However, only a fraction of the pulp produced in South Africa is used locally.

The forest industry stores a total of 3.69 Tg C yr^{-1}, which is equivalent to 13.5 Tg CO_2, or about 3.8% of the annual CO_2 emissions for South Africa. Although not insignificant, this amount is too small to make a serious impact on the national emission budget. The potential for increasing C storage via this avenue in South Africa is very small because of the adverse environmental impacts of further plantations and the limited area that is bioclimatically suitable for afforestation (Winjum *et al.*, 1993). The potential in wetter African countries is greater, but only if the plantations are established in areas with a low existing C density. The clearing of endemic woodlands for replacement with pines and eucalypts will not necessarily meet this criterion.

4. Conclusions

The methods for calculating C sequestration through afforestation are relatively simple to apply, but obtaining reliable data on biomass C densities and rates of afforestation may be difficult in some countries. Data on nontimber biomass fractions, especially belowground, are generally sparse. The C density and growth data used in this analysis for South Africa are applicable for all of southern Africa.

The use of peak biomass for the calculation of C density is not appropriate for rotation forestry. A mean C storage approach, which averages the C density over the rotation, is recommended (Winjum *et al.*, 1992; Winjum *et al.*, 1993). The dynamic method of C storage calculation, which accounts for the age of various plantations, is recommended where the rate of afforestation varies between years.

Storage of C in long-lived forest products is as important as new afforestation as a C sink in countries where the national pool of timber products is growing, and should be considered as part of the emission inventory. Additional research is needed on changes in C storage in the soil and litter layers and the fate of paper products.

References

Christie, S.I.: 1991, *The Biomass Production of* Eucalyptus grandis *Planted at Various Close Spacings*, FOR-DEA 288, Division of Forest Science, CSIR, Pretoria.

Brown, S., and Lugo, A.E.: 1982, The storage and production of organic matter in tropical forests and their role in the global carbon cycle, *Biotropica*, **14**, 161–187.

Dixon, R.K., Winjum, J.K., Andrasko, K.J., Lee, J.L., and Schroeder, P.E.: 1994, Integrated land use systems: assessment of promising agroforest and alternative land use practices to enhance carbon conservation and sequestration, *Climatic Change*, **27**, 71–92.

Everson, T.M., van Wilgen, B.W., and Everson, C.S.: 1988, Adaptation of a model for rating fire danger in the Natal Drakensberg, *South African Journal of Science*, **84**, 44–49.

FOA (Forest Owner's Association): 1994, *South African Forestry Facts*, Forest Owner's Association, Rivonia, South Africa.

Gerischer, G.F.R: 1993, Pulp and paper manufacture in South Africa, in: van der Sijde, H.A. (ed.), *Forestry Handbook*, The South African Institute of Forestry, Pretoria, 498–519.

Harmon, M.E., Ferrell, W.K., and Franklin, J.F.: 1990, Effects of carbon storage of conversion old-growth forests to young forests, *Science*, **240**, 699–702.

Harrison, W.M., Pienaar, L.V., and Kotze, H.: 1994, *Planting Density Effects on Growth and Yield in the Nelshoogte Correlated Curve Trend Experiment with* Pinus patula, CSIR report no. FOR–C–264, 71 pages.

Herbert, M.A., and Robertson, M.A.: 1991, *Nutrient Distribution, Cycling and Loss Within Stands of Commercial Timber Species*, Annual Research Report, Institute for Commercial Forestry Research, Pietermaritzburg, South Africa, 54–59.

IPCC/OECD (Intergovernmental Panel on Climate Change/Organisation for Economic Co-operation and Development) Joint Programme: 1994, IPCC Draft Guidelines for National Greenhouse Gas Inventories, IPCC/OECD Joint Programme, Paris, 3 volumes.

Kinerson, R.S., Ralston, C.W., and Wells, C.G., 1977, Carbon cycling in a loblolly pine plantation, *Oecologica*, **29**, 1–10.

Kotze, H.: 1991, *Growth and Yield Prediction Methodologies—An Implementation of Different Strategies*, forest management seminar, Univ. Stellenbosch, 29 pages.

Loveday, N.C., and Kassier, H.W.: 1993, Yield tables for some of the major timber species grown in South Africa, in van der Sijde, H.A. (ed.), *Forestry Handbook*, The South African Institute of Forestry, Pretoria, 303–323.

LHA Management Consultants: 1993, *The Current and Forecast Demand for Roundwood in South Africa*, report to SATGA, Pretoria.

Morris, A.R.: 1992, Dry matter and nutrients in the biomass of an age series of *Pinus patula* plantations in the Usutu forest, Swaziland, *SA Forestry Journal*, **163**, 5–11.

OECD (Organisation for Economic Co-operation and Development): 1991, *Estimation of Greenhouse Gas Emissions and Sinks*, Organisation of Economic Cooperation and Development, Paris.

Schroeder, P.E.: 1992, Carbon storage potential of short rotation tropical tree plantations, *Forest Ecology and Management*, **50**, p 31–41.

UNEP, OECD, IEA, IPCC (United Nations Environment Programme, Organisation for Economic Co-operation and Development, International Energy Agency, Intergovernmental Panel on Climate Change): 1995, *IPCC Guidelines for National Greenhouse Gas Inventories*, IPCC, Bracknell, 3 Volumes.

van Vuuren, N.J.J., Banks, C.H., and Stohr, H.P.: 1978, *Shrinkage and Density of Timbers Used in the Republic of South Africa*, Bulletin 57, Department of Forestry, Pretoria.

Winjum, J.K., Dixon, R.K., and Schroeder, P.E.: 1992, Estimating the global potential of forest and agroforest management practices to sequester carbon, *Water, Air and Soil Pollution*, **64**, 213–227.

Winjum, J.K., Dixon, R.K., and Schroeder, P.E.: 1993, Forest management and carbon storage: an analysis of 12 key forest nations, *Water, Air and Soil Pollution*, **70**, 239–257.

THE BURNING OF FUELWOOD IN SOUTH AFRICA: WHEN IS IT SUSTAINABLE?

G.P. VON MALTITZ and R.J. SCHOLES

Forestek, CSIR, P.O. Box 395, Pretoria 0001, South Africa

Abstract. Fuelwood harvesting is considered sustainable when consumption is equal to or less than production. An empirical model was developed to estimate potential fuelwood production from savannas. The model is based on the observation that in semiarid savannas, biomass production is linearly dependent on rainfall. Woody basal area is linked to mean annual precipitation, and aboveground woody biomass is proportional to basal area. Production averages 4% of standing woody biomass and is corrected to exclude stems that are too small for harvesting. The model assumes that the entire area consists of seminatural savanna. Corrections for the land lost as a result of land transformation and degradation would have to be included. Data on land loss can most effectively be obtained from satellite imagery, with appropriate ground calibration. The model is based on limited data sets but in most instances has been validated against independently collected data. The model yields a reasonable prediction at a national and regional level, but estimates for limited areas or specific points on the ground may differ substantially from the predicted values. Model results indicate a potential for sustainable fuelwood production at the national level, but specific regions are using fuelwood at nonsustainable levels.

1. Introduction

The Intergovernmental Panel on Climate Change (IPCC) emissions inventory methods consider unsustainable use of fuelwood to be a source of net carbon dioxide (CO_2) emissions (UNEP *et al.*, 1995). Completion of the inventory, therefore, requires an estimation of the amount of fuelwood that is used unsustainably, which in turn requires a consideration of both consumption and production patterns.

The burning of fuelwood is still the primary source of energy for most of sub-Sahara Africa (Anderson, 1986)—the majority of the rural population, even in relatively developed countries such as South Africa, still use wood as their primary energy source. Numerous studies indicate that in developing countries, wood is being consumed at nonsustainable levels (Bajracharya, 1983; Allen, 1985; Shackelton, 1994; Kaul, 1993; Kanetkar and Kaul 1994). Within South Africa, it is estimated that approximately 11 million metric tons of fuelwood is consumed annually (Gander, 1994). However, in South Africa there are no data based on detailed empirical studies to use in estimating the size of the sustainable fuelwood resources in the extensive woodland areas of the country.

Accurate estimates of fuelwood consumption are difficult because of methodological constraints and unreliable population demographics data. A number of estimates, largely based on questionnaires but with a few based on actual measurement, have been conducted within South Africa (summarized in Gander, 1994). Annual consumption rates vary greatly, from less than 250 kg per capita (Eberhard and Dickson, 1987) to more than 1100 kg per

Environmental Monitoring and Assessment **38**: 243-251, 1995.

capita (Best, 1979) in rural areas and from 26 kg per capita to more than 742 kg per capita in semirural/periurban areas (Eberhard, 1986). Workers on commercial farms are also often dependent on fuelwood—the few available studies estimate 800 to 900 kg per capita (Gander, 1992). Gander (1994) estimates total fuelwood consumption for South Africa to be 11.3 million megagrams (Mg), with the bulk, 6.6 million Mg, consumed in the ex-Bantustans and an additional 3.5 million Mg consumed by farm workers. Wood consumption rates depend on numerous issues such as wood availability, cost and availability of alternative fuels, levels of poverty, and the severity of the need for space heating. Not all fuelwood consumed is from the savanna regions. Forest patches, small woodlots, and wattle groves (*Acacia* spp.) are important wood sources, particularly in the grassland biome. Waste from commercial forestry also contributes to fuelwood availability but is probably of relatively limited importance at present (Gander, 1994).

The estimation of savanna primary production is based on the observation that in semiarid savannas, the relationship between biomass production and rainfall is linear (Scholes, 1990, 1993). Although the assumption of linearity is robust, site-specific circumstances are such that the relationship between biomass production and rainfall for any particular site is in some ways unique. For instance, the depth, texture, and nutrient status of the soil affect both the slope and intercept of the relationship. Factors such as temperature and light also influence primary production because they influence the physiological processes of the savanna vegetation. The generalized relationships developed in the following sections are based on a large number of sites encompassing a wide range of variation. For any particular site, the prediction of biomass is likely to be erroneous, but for regional averages involving many sites, the central tendency is both robust and sufficiently accurate for the purpose of broad survey.

The objective of this paper is to present a simple method for estimating sustainable biomass production on a regional scale, and to discuss the suitability of using this data as a basis for estimating CO_2 emissions from unsustainable fuelwood burning.

2. Method

Mean annual precipitation was chosen as a suitable independent variable for nationwide predictive models, because it is available for all of Africa at a 30 × 30 m resolution and for South Africa at a 1 × 1 m resolution from the Computer Centre for Water Research (CCWR). Temperature data at a similar resolution are not currently available for the country, although they are being developed. Soil data are extremely difficult to use in national predictive models because of the spatial variability in soil properties at the local scale, the unavailability of a national digital database of soil properties, and a lack of understanding about the ways in which different soil properties affect tree growth.

Very few studies have investigated either total aboveground tree biomass or tree production rates within the savannas of southern Africa. It is, therefore, not possible to develop a direct relationship between either of these two parameters and mean annual precipitation. The following four-step approach was used to predict woody production potentials:

In step 1, a linear relationship was developed between the basal area of woody vegetation (*BA*: m^2/ha) and the mean annual precipitation in South African savannas (*MAP*: mm). Forty-three sample quadrats from fourteen sites spread throughout the savanna region of South Africa were used to develop this relation, and twenty plots from an independent study were used to validate it. Study sites were located in areas of seminatural savanna, and sample quadrats were located at a range of catenal positions. Basal area was calculated as the summation of the individual tree-stem area derived from stem-diameter measurements. Although there was a large degree of variance within the data, a significant correlation between basal area and rainfall was established ($r^2 = 0.29$, $p < 0.05$). The regression equation derived ($BA = -0.03535 + 0.0158\ MAP$) was very similar to a regression derived by Frost, of the Department of Biological Sciences, University of Zimbabwe, from a totally independent data set collected in Central Africa. The linear model based on South African data begins to deviate from Frost's exponential model only when rainfall values are greater than 1200 mm (see Figure 1). These rainfall values are greater than the expected rainfall in almost all South African savannas. Maximum basal area was limited to 32 m^2/ha, which represents the upper limit of the data.

In step 2, the aboveground woody biomass (*WB*:t/ha) was related to stand basal area. A linear regression model linking woody basal area to aboveground woody biomass was derived from Southern African data sources (Moore *et al.*, 1967; Scholes, 1988; Rutherford, 1982; Endean, 1967; Dayton, 1978). Although only seven data points were available, a highly significant regression equation ($WB = -2.930 + 2.7\ BA$) was fitted ($r^2 = 0.93$, $p < 0.001$). This equation compares favorably to Frost's exponential model derived for basal area values less than 10 m^2/ha. At higher values, the linear model gives more conservative results. Maximum aboveground biomass was limited to values less than 144 t/ha based on data from Malaisse and Strand (1973).

In step, 3, the annual basal area increment was related to the stand basal area. Very few data sets exist for this purpose (Rutherford, 1984; Lubke and Thatcher, 1983; Rutherford and Kelly, 1978; van Vegten, 1984; Malaisse *et al.*, 1975; Endean, 1967). The values range from 2.1% to 9.4%, with a mode around 4% (Scholes and Walker, 1993). These data are often unreliable because of the complexities involved in measuring stem increment, and because the data were often collected over too short a span of years to adequately represent the fluctuations in rainfall. There appears to be a trend from higher values (6%) at sites with low basal area, to low values (2%) at sites with high basal area, but this trend is not statistically significant. The modal value used was 4%, which corresponds to the rainfall range in South African sites, if indeed this relationship exists. This value represents the weakest point in the procedure, and could be in error by ± 50% because of the range in published values.

In step 4, a correction was made for that fraction of the aboveground biomass that is not considered useful as fuel. The lower size limit for fuel is an arbitrary decision, because

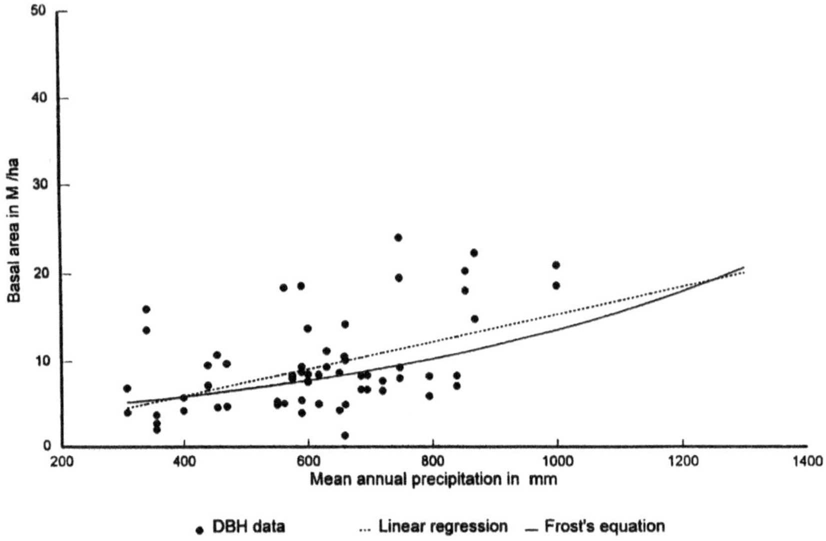

Figure 1. DBH (diameter at breast height) data from sixty-two sample plots showing the fitted regression equation and Frost's exponential model derived from independent data.

it varies with the use pressure. Output indicates that the model is not very sensitive to the influence of rainfall on the correction factor. A value of 0.025 m for the twig diameter (*TD*:m) was used. The approach is based on the "pipe theory" for tree architecture, which predicts that the total cross-sectional area of a tree is relatively constant from the base to the twigs. Guy (1981) presents an equation that relates stem diameter to wood mass for individual savanna stems. The relationship appears stable over the entire data range. The wood biomass for the whole stem is, therefore, calculated by applying the equation to the mean stem (*MSD*:m) diameter at the site ($WB = WB$ (1.0 – exp (log (TD/MSD) 0.5101)). The mass of an equivalent cross-sectional area, made up of 25 mm diameter twigs, is then subtracted.

The mean stem diameter at a site is derived from the observation that savannas across a broad rainfall range tend to have 500 to 1000 (averaging 800) stems per hectare that are greater than 50 mm in diameter. Substitution of the Guy relation into the predicted biomass per tree allows the mean stem diameter to be estimated as ($MSD = 0.1807$ exp(0.00019604 MAP)). It is a weak increasing function of rainfall. The result of these calculations reduces the predicted total woody biomass by 22.5% to 25% to yield a usable woody biomass estimate.

The data are applicable only to savanna areas. A digitized and rasterized vegetation map (Acocks, 1988) was used to determine the extent of the savanna biome in South Africa.

The classification criteria of Rutherford and Westfall (1986), with modifications suggested by Ellery (1992), were used to reclassify the map to represent biomes. Data analysis was conducted with the CCWR rainfall data, a rasterized map of the Bantustans, and a derived biome map as input. The calculations were made with custom code written in the BASIC programming language on a microcomputer, and the resulting output was generated with IDRISI software. Results were computed at the national level, and for each ex-Bantustan for comparison against published consumption levels.

3. Results

A map of the predicted wood production in South African savannas is shown in Figure 2. The predicted total annual production of usable fuelwood from savannas in South Africa is 20 million Mg, assuming no land-use transformation or land degradation. This estimate is based on a predicted total standing woody biomass of 762 million Mg.

Table I shows the model's predictions of potential wood production from the savanna regions of the Bantustans. This is compared to published estimates of consumption per Bantustan.

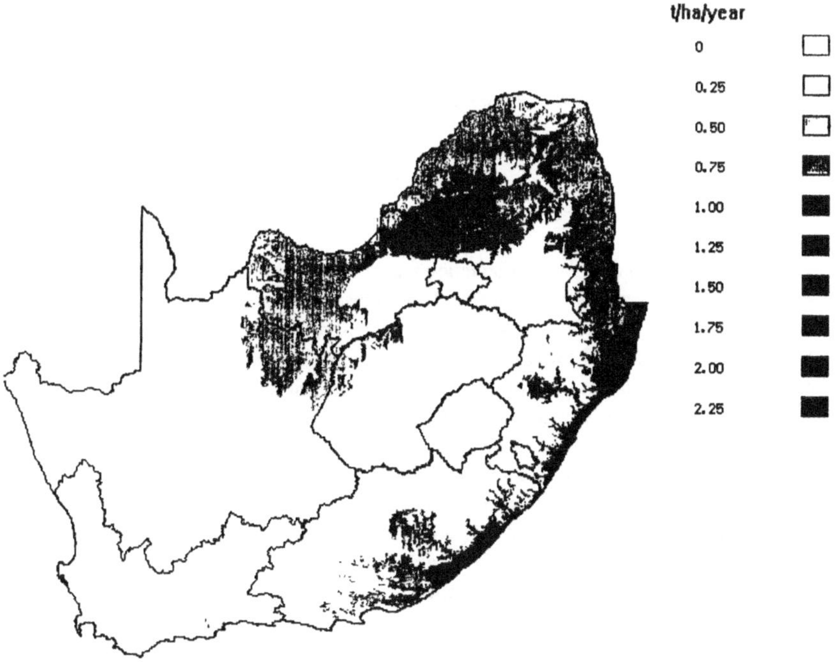

Figure 2. Potential biomass production of South African savannas in Mg ha yr.

TABLE I

Predicted fuelwood consumption rates versus potential production in the ex-Bantustan areas.

Bantustan	% Savanna	Annual Consumption (000 Mg)*	Factored Consumption (000 Mg)	Potential Savanna Production (000 Mg)
Bophuthatswana	93.5	483	451	1494
Ciskei	69.7	193	135	310
Gazankulu	99.9	452	452	362
KaNgwane	73.3	283	72	165
KwaNdebele	91.7	98	90	359
KwaZulu	63.9	2193	1401	1652
Lebowa	85.5	1271	1086	970
QwaQwa	0	26	0	0
Transkei	24.6	1311	322	1070
Venda	75.7	315	238	249

* From Gander, 1994

Note: Consumption has been factored by the percentage of the area under savanna. This is a conservative estimate because fuelwood is likely to be transported out of the savanna areas, and consumption is likely to be highest in the savanna areas where fuelwood is more easily obtainable.

4. Discussion

The predicted sustainable wood production rates presume the existence of healthy savanna vegetation over the entire savanna region. Land transformation or degradation has, however, accrued over most of the savanna region. Analysis of satellite imagery could provide a cost-effective mechanism for correcting for land transformation and degradation, but this would require field calibration of the effects of different levels of observed degradation on standing basal area. Up to this point, imagery analysis has been conducted only on limited areas and with limited or no ground verification.

The results indicate that at a national level, there is a potential for sustainable fuel consumption, assuming that most of the savanna areas are maintained in a seminatural state. This would mean zero net CO_2 emissions, given uniform usage of the resource. However, it must also be noted that fuelwood consumption is not uniform over the savanna regions; high nonsustainable consumption is concentrated around areas with high population densities. Within South Africa and many other African states, betterment programs (resettlement programs forcing communities into villages) have resulted in high-density nuclear settlements. These settlements are often located close to agricultural fields that are totally cleared of bush, which reduces the area available for fuelwood production.

Table I indicates that in some of the ex-Bantustan areas there is potential for sustainable fuel production, but this conclusion does not take into account the fact that land degradation and transformation have greatly reduced the area available for production of fuelwood. In other ex-Bantustans, consumption exceeds maximum potential production. Fuelwood consumption is probably at unsustainable levels at localized sites in all the ex-Bantustans, although there is potentially a surplus in a few instances. Harvesting fuelwood for sale in nonsavanna regions and periurban and urban areas may further increase the level of consumption in some regions.

Although South Africa has almost no charcoal trade, in nations such as Tanzania, Malawi, Zambia, and Mozambique charcoal trade is very active. Once areas of woodland have been cleared for charcoal production, they are often converted to agricultural use, and no tree regeneration can occur.

Fuelwood consumption patterns change as fuelwood becomes scarcer. Traditional resource management practices on communal land limit fuelwood collection to dead wood; and in such cases, fuelwood consumption is sustainable. However, increased population pressure and the conversion of land from woodlands to agriculture or other uses increase the pressure on the remaining woodlands. This can lead to the harvesting of live wood to meet the shortfall in fuelwood. If this occurs, there will be a net emission of CO_2 from the region. However, as fuelwood becomes scarce, its value increases, which often results in a switch to alternative energy sources, such as paraffin. The system may then stabilize at a lower level of sustainable wood consumption, or the wood resources may become locally depleted.

4.1 Precautions when Interpreting the Data

Because this is the first time that a production (versus a standing crop) estimate has been made for savannas in South Africa, there are no data to verify the prediction. It should, therefore, be regarded as tentative.

Each step in the modeling process relies on regression equations. The data on which these models are based are, in most instances, limited to fairly small data sets. Where possible, these have been validated against independently collected data.

The models should give a reasonable prediction of total woody biomass production of savannas at the national level, but estimates for limited areas or specific points on the ground may differ substantially from the values predicted in the model. The results should not be expressed at a scale smaller than the magisterial district.

5. Conclusions

The methodology described in this paper provides an efficient way to predict potential fuelwood production at a regional, national, or continental level. At a regional level predictions about fuelwood production can be combined with fuelwood consumption data to allow for calculation of unsustainable consumption. It must be recognized, however, that consumption patterns are spatially heterogeneous and that localized overconsumption is not compensated for by underconsumption in other areas. Although there is potential for sustainable consumption in South Africa at a national level, localized consumption is not operating in a sustainable fashion, and this is leading to net CO_2 emission and degradation of the savanna resource.

References

Acocks, J.P.H.: 1988, Veld types of south Africa. *Memoirs of the Botanical Survey of South Africa* **57**, 146 pages.

Allen, J.C.: 1985, Wood energy and preservation of woodlands in semi-arid developing countries. The case of Dodoma region, Tanzania, *Journal of Development Economics* **19**, 59–84.

Anderson, D.: 1986, Declining tree stocks in African Countries, *World Development* **14**, 853–863.

Bajracharya, D.: 1983, Fuel, food or forest? Dilemmas in a Nepali village, *World Development* **11**, 1057–1074.

Best, M.: 1979, The scarcity of domestic energy: a study of three villages, SALDRU working paper no. 27, University of Cape Town.

Dayton, B.R.: 1978, Standing crops of dominant *Combretum* species at three browsing levels in the Kruger National Park, *Koedoe* **21**, 67–76.

Eberhard, A.A.: 1986, *Energy Consumption Patterns in Underdeveloped Areas in South Africa*, Energy Research Institute Report No 94, University of Cape Town, South Africa.

Eberhard, A.A. and Dickson, B.J: 1987, *Energy Consumption Patterns and Alternative Energy Supply Strategies for Underdeveloped Areas of Bophuthatswana*, Energy Research Institute, University of Cape Town, South Africa.

Ellery, W.N.: 1992, *Classification of the Vegetation of the South African Grassland Biome*, PhD thesis, University of the Witwatersrand, Johannesburg, South Africa, 211 pages.

Endean, F.: 1967, The productivity of miombo woodland in Zambia, *Forest Research Bulletin* **14**, Lusaka, Zambia: Government Printer.

Frost, P.G.H: Undated, The ecology of miombo woodlands. Unpublished manuscript in preparation. Department of Biological Sciences, University of Zimbabwe, Harare, Zimbabwe.

Guy, P.R.: 1981, The estimation of the aboveground biomass of the trees and shrubs in the Sengwa Wildlife Research Area, Zimbabwe, *Southern African Journal of Wildlife Research* **11**, 135–142.

Gander, M.: 1992, *Domestic Energy Used by Farmworkers Living on Commercial Farmland in Natal and Transvaal*, Report to the Department of Mineral and Energy Affairs, South Africa.

Gander, M.: 1994, *Afforestation and Woodland Management in South Africa*. EDRC paper no 9, University of Cape Town, South Africa.

Lubke, R.A. and Thatcher, F.M.: 1983, Short term change in the Burkea savanna, *South African Journal of Botany* **2**, 85–97

Kanetkar, R.S and Kaul. O.N.: 1994, *The energy dimension of Indian Forests*. Workshop of the joint research project on the forests in South- and North-transition from deforestation to sustainable forest policies in redressing global warming, Joensuu, Finland, 17 pages.

Kaul, O.N.: 1993, Forest biomass burning in India, in *The Climate Change Agenda: An Indian Perspective*, 65–90, Tata Energy Research Institute, New Delhi.

Malaisse, F., Freson, R., Geoffinet, G., and Maalaisse-Mousset, M.: 1975, Litterfall and litter breakdown in miombo, in Golley, F.B. and Medina, E. (eds.), *Tropical Ecological Systems: Trends in Terrestrial and Aquatic Research*, Berlin: Springer-Verlag, 137–55.

Malaisse, F. and Strand, M.A.: 1973, A preliminary miombo forest seasonal model, in Kern, L. (ed.), *Modeling Forest Ecosystems*, I.B.P., Woodland Workshop, Oak Ridge, USA, 29–295.

Moore, A.W., Russel, J.S., and Coaldrake, J.E.: 1967, Dry matter and nutrient content of a subtropical semiarid forest of *Acacia harpophylla* F. Muell. (Bigalow). *Australian Journal of Botany* **15**, 11–24.

Rutherford, M.C. and Kelly, R.D.: 1978, Woody plant basal area and stem increment in *Burke africana*—Ochna pulchra woodland, *South African Journal of Science* **74**, 307–308.

Rutherford, M.C.: 1982, Woody Plant Biomass Distribution in *Burkea africana* Savannas, in Huntley, B.J. and Walker, B.H. (eds.), *Ecology of Tropical Savannas. Ecological Studies* **42**, Berlin: Springer-Verlag.

Rutherford, M.C. and Westfall, R.H.: 1986, Biomes of southern African objective categorization, *Memoirs of the Botanical Survey of South Africa*, **54**.

Rutherford, M.C.: 1984, Relative allocation and seasonal growth of woody plant components in a South African Savanna, *Progress in Biometeorology* **3**, 45–63.
Shackelton, C.M.: 1994, *Sustainable Veld Management and Its Role in Social Forestry and Energy Planning*. Plant for Life Conference Proceedings, September 1994, Department of Mineral Affairs and Forestry, Pretoria, South Africa, 24 pages.
Scholes, R.J.: 1988, *Response of Three Semi-Arid Savannas on Contrasting Soils to the Removal of the Woody Component*, PhD Thesis, University of the Witwatersrand, Johannesburg, 297 pages.
Scholes, R.J.: 1990, The influence of soil fertility on the ecology of southern African dry savannas, *Journal of Biogeography* **17**, 415–419.
Scholes, R.J.: 1993, *Nutrient Cycling in Semi-Arid Grasslands and Savannas: Its Influence on Pattern, Productivity and Stability*, Proceedings of the XVII International Grassland Congress, 1993.
Scholes, R.J and Walker, B.H.: 1993, *An African savanna: synthesis of the Nylsvley study*, Cambridge: Cambridge University Press.
UNEP, OECD, IEA, IPCC (United Nations Environment Programme, Organisation for Economic Co-operation and Development, International Energy Agency, Intergovernmental Panel on Climate Change): 1995, *IPCC Guidelines for National Greenhouse Gas Inventories*, IPCC, Bracknell, 3 Volumes.
van Vegten, J.A.: 1984, Thornbush invasion in a savanna ecosystem in eastern Botswana, *Vegetation* **56**, 3–7.

BIOFUEL USE ASSESSMENTS IN AFRICA: IMPLICATIONS FOR GREENHOUSE GAS EMISSIONS AND MITIGATION STRATEGIES

D.L. KGATHI[1] and P. ZHOU[2]

[1]*National Institute of Development Research and Documentation, University of Botswana, Private Bag 0022, Gaborone, Botswana.* [2]*Energy, Environment, Computer and Geophysical Applications Consultants, P.O. Box 402339, Gaborone, Botswana.*

Abstract. The energy balances of most African countries suggest that biofuels (woodfuel, crop and wood residues, and dung) constitute the largest share of total energy consumption (up to 97% in some sub-Saharan Africa countries). There is, however, an increasing scarcity of woodfuel (fuelwood and charcoal), the major biofuel, and a feared increase in greenhouse gas (GHG) emissions associated with biofuel combustion. The extent of GHG emissions is estimated from biofuel consumption levels that are in turn based on methodologies that might be inaccurate. A questionnaire, supplemented by informal interviews, are used to collect data, yielding information regarding end-uses, technologies used, scale of consumption, determinants of fuel consumption, and interfuel substitution (among other parameters). The survey revealed that cooking is the major end-use, with other common uses, such as space and water heating. Improved stoves that provide better combustion efficiency and, thus, reduce woodfuel consumption have not been widely disseminated and are associated with higher methane emissions than open fires. More than 90% of the households in Africa use open fires. Consumption is presented as per capita for households and as products and quantity of fuel in the small scale industries, commercial, and public sectors. Among the determinants for biofuel consumption are affordability, availability of the fuel, and interfuel substitutions. Flaws in estimating biofuel consumption yield large uncertainties in GHG emissions, with implications for the development of policies on energy planning and environmental protection. However, the application of scenarios can guide policy formulation.

1. Introduction

Biomass is an important source of energy for the majority of the population in developing countries (Agarwal, 1986). In sub-Saharan Africa (excluding South Africa), it accounts for 73% of the total energy consumption (Table I). In Burkina Faso, Ethiopia, Malawi, Rwanda, Somalia, Tanzania, and Uganda, biomass accounts for more than 80% of total energy consumption. Only a few countries, such as Zimbabwe, Seychelles, and Mauritius, depend on biomass fuels for less than 50% of total energy consumption (Table II). The most important sources of biomass fuels in Africa are woodfuel (fuelwood and charcoal), crop and wood residues, and dung (Hall, 1994). Woodfuel is, undoubtedly, the most important source of biomass energy and, because of its increasing scarcity, it is a subject of major concern in Africa and the rest of the developing world.

The bulk of biomass is consumed by the domestic sector, whereas a small, but significant proportion is consumed by small-scale industrial and commercial enterprises and public institutions (Eberhard, 1986). The combustion of biomass is associated with the production of GHGs, namely carbon dioxide (CO_2), methane (CH_4), nitrous oxides (N_2O), nitrogen oxides (NO_x), and ozone (O_3). In the *IPCC Guidelines for National Greenhouse Gas*

Environmental Monitoring and Assessment **38**: 253-269, 1995.

TABLE I

Estimated energy consumption in Africa by region (1990)

	North Africa		Sub-Saharan excluding Republic of South Africa		Republic of South Africa	
Population (million)	114	(17%)	497	(77%)	39	(6%)
Fuel	Mill TOE	EJ TOE	Mill	EJ TOE	Mill	EJ
Non-Biomass Fuels	80	3.4	47	2.0	83	3.6
Biomass Fuels	10	0.4	126	5.4	4	0.2
Total	90	3.8	173	7.4	87	3.8
		(26%)		(49%)		(25%)

Source: Ardayfio-Schandorf, E.

TABLE II

Biomass consumption as % of total energy in selected African countries

Country	Biomass as % of Total Energy	Country	Biomass as % of Total Energy
Botswana	58	Rwanda	97
Burundi	95	Senegal	61
Egypt	28	Seychelles	9
Ethiopia	94	Somalia	86
Gambia	75	Sudan	95
Ghana	63	Tanzania	97
Kenya	86	Tunisia	27
Mauritius	46	Uganda	95
Morocco	13	Zaire	86
Mozambique	94	Zambia	63
Nigeria	82	Zimbabwe	43

Source: Hall, D.O. and Rossilo-Calle, F. (1991)

Inventories (UNEP *et al.*, 1995), CO_2 is generally considered to be recirculated in the biosphere system. For the purposes of policy analysis, all GHG emissions from biofuel combustion are considered significant.

To develop greenhouse gas emission inventories and mitigation strategies, it is essential to know the patterns of biofuel use in Africa (Brokensha and Castro, 1993). A number of surveys on biomass use in Africa have been undertaken, but there is concern that some of the methods used have flaws. Also, the results of the various surveys are not always known to policy makers, because they are scattered in a number of reports (Kgathi, 1990).

This paper reviews the various biomass energy surveys conducted in Africa (Tomkins, 1986). More specifically, it examines the methods used in biomass surveys including the type of interview and biomass measurement methods used. It also examines the results of surveys pertaining to end-uses of biomass, technologies used for burning biomass, magnitude and determinants of biomass consumption, and the patterns of interfuel substitution. Finally, the implications of the patterns of biomass use for GHG emissions and related climate change issues are presented.

2. Methods of data collection in households

This section reviews methods used in conducting bioenergy household surveys and describes the sampling techniques used. Methods of data collection include the structured questionnaire, informal interviews, group interviews, observations, and physical measurement of biomass (Kaale, 1986).

2.1. Bioenergy Studies and Sampling Methods

Most of the bioenergy consumption studies conducted in Africa have focused on woodfuel, rather than other biofuels such as crop residues and dung (Hall, 1994). At one extreme, there are national studies to collect data that is needed for the formulation of energy policy and planning. As Mwandosya (1986) points out, "These surveys do not need to be of high intensity, and usually use rather simple data gathering methods," and, consequently, often lack detail. These surveys may also collect nonenergy data, although such data collection is often limited. Kenya is an example of a country in which extensive national surveys have been conducted by the government, the Beijer Institute, and USAID (Hosier and Dowd, 1988; FAO, 1983).

At the other extreme, there are microsurveys, which are surveys undertaken at the level of a locality (village or town). Microsurveys are often intensive and research oriented (Kgathi, 1992). They involve a broad range of questions, including questions on socioeconomic issues. As a result, these surveys tend to highlight the inter-relationships between energy and nonenergy variables, but because they focus on one small part of the country, they lack coverage and fail to address a diversity of issues (ESMAP, 1991).

Most of the bioenergy surveys in Africa have used probability sampling, a technique whereby the sample is randomly selected but representative of the composition of the population in the country (Leach and Mearns, 1988). Sampling is often guided by some existing framework, such as population census or previous surveys. Where there is no sampling framework, some stratification system can be developed or cluster sampling is used. For instance, the Malawi rural energy survey undertaken in the 1980s used the sampling frame of the national agricultural surveys (FAO, 1983). Also, in a microstudy undertaken in Sudan, Digernes (1977) used local rate lists as a sample framework to construct a stratified random sample. To conduct a study on energy use (fuelwood and other energy sources) in Kenya, Hosier (1985) used the cluster sample frame known as the National Integrated Household Survey. The sample was based on ecological zones. Finally, fuelwood surveys undertaken by Kgathi (1992) and Kgathi *et al.* (1994) in Botswana also relied on a frame of the past census to select a two-stage, stratified random sample.

2.2. The Structured Questionnaire and Other Methods

The questionnaire is the most frequently used method for collecting data on biofuel use both in low latitude nations. Questionnaires on biofuel use in Africa usually include a range of questions on such aspects as end-uses of the fuels, technologies used in burning the fuels, patterns of interfuel substitution, and the demography and incomes of households (Kgathi, 1992).

The questionnaire method has a number of advantages. It allows the results to be generalized because uniform questions are asked, and it can be administered by enumerators when researchers are absent (Bhatia, 1985; Tomkins, 1986). However, complete reliance on the questionnaire may not be rewarding at times, particularly if the issue being investigated is a sensitive one. Consequently, most researchers on bioenergy use in Africa have supplemented the questionnaire method with other survey techniques such as the informal interview, observations, and group interviews. For instance, in studies on fuelwood use undertaken by Kamara (1986) and Davidson (1985) in Sierra Leone and Kgathi (1992) in Botswana, the structured questionnaire was supplemented with informal open-ended interviews in an effort to obtain additional data not available through precoded questionnaires. One of the advantages of the informal interview is that it enables the researcher to check the reliability of information collected by the questionnaire. This technique, known as triangulation (Abel *et al.*, 1988), is very effective in improving the quality of data.

The observation method can either be participant or nonparticipant (Karenzi, 1994). The former technique is often used by anthropologists, and it requires the researcher to become part of the study population. The main problem with this method is that it is too demanding and time consuming. Consequently the method is rarely used. According to Kgathi (1990), only one survey used participant observation in Botswana to collect data on fuelwood. The researcher lived in Oodi, which was one of the study areas, and collected data on various aspects of fuelwood procurement and use. In nonparticipant observation, the researcher does not become part of the study population. Nonparticipant observation is the method most frequently used in bioenergy studies in Africa. It has been used to collect such data as the types of fuels used and the technologies used for burning biomass in Botswana (Gay and Zietlow, 1993, Kgathi, 1990).

2.3. Biomass Measurements

Two primary methods are used to determine the rates of household fuelwood consumption in Africa. One method is based on the respondents' own estimates of biofuels consumed—referred to as the user-recall method (Tomkins, 1986); the other is based on the actual amount of fuel consumed (Eberhard, 1986; Hosier, 1985). In the user-recall method, households are asked to estimate the amount of biomass they use per period of time (often per day). This method has been used in most countries to estimate household consumption. In Africa, the user-recall method has been used in a number of surveys in Kenya, South Africa, Botswana, and Zimbabwe (Eberhard, 1986; Hosier, 1985; Tietema *et al.*, 1988; Zhou, 1994). However, one of the limitations of the user-recall method is its reliance on the interviewee's memory; the method is subjective and can produce unpredictable, serious errors. The main advantage is the speed at which the information on biomass consumption can be collected.

In the physical measurement method, households are asked to put aside the amount of fuelwood they are likely to use for a certain period, such as five days. After three days, the remaining fuelwood is weighed to estimate the actual amount consumed by the household. The researcher must ensure that the fuelwood has actually been used by the household and not borrowed by or given to neighbors or friends or that no new fuelwood has been added to the pile. The physical measurement approach can yield more accurate results if it is done repetitively and if the respondents are reliable (Brokensha and Castro, 1994). It is time consuming only in that it entails repeated measurements at the same households for a number of months.

The accuracy of the methods for estimating the rates of fuelwood consumption was assessed in Botswana and Rwanda by researchers of the African Energy Policy Research Network (AFREPREN) (Kgathi and Mlotshwa, 1994). In Botswana, repetitive measurements of fuelwood were taken by enumerators for six months in the village of Khakhea, nine months in Mmankgodi, and three months in Molepolole. The weight of the fuelwood found in each visit was recorded, and the difference between the totals from one day to the next was the daily household fuelwood consumption. The data on the daily household fuelwood consumption was compiled over the period stated above, then the mean daily household fuelwood consumption was estimated.

The same procedure was followed in Rwanda, where repetitive measurements of bioenergy were taken for one year, with the exception of one region where the measurements were taken for 10 months. Unlike the Botswana assessment, biofuel measurements in Rwanda were not limited to fuelwood but also included residues and dung (Karenzi, 1994).

Another method of data collection is monitoring headloading and carts along fuelwood collection routes and measuring the quantity of the fuelwood transported and frequency of collection (Mwandoysa, 1986). The difficulty with this method is that it is impossible to guarantee that the fuelwood is consumed in one household, and the sizes of headloads and carts cannot be generalized. Consumption rates derived from this method will obviously contain larger errors than the rates derived from the other two methods, but comparative figures are not available.

In all the above methods, household consumption is converted to per capita consumption by dividing the household consumption figure by the number of members in the household. Unfortunately, researchers do not specify what they consider household size to be. It might be necessary to convert household members to adult-equivalents. Per capita consumption is also not very sensitive to household size, because an additional member may not invoke extra cooking or space heating.

2.4. Data Analysis

Most of the biomass surveys in Africa use simple and descriptive statistical techniques in data analysis, such as measures of central tendency, measures of dispersion, bar charts, and simple tables (Kgathi, 1989). More sophisticated techniques such as regression analysis, chi square tests, and analysis of variance are rarely used. Considerable bioenergy data can be derived from survey information. For instance, price and income elasticities of the demand for biofuels could be estimated, if regression analysis were used. However, lack of reliable time series or spatial data sometimes prohibit this approach.

3. Results of assessments

3.1. Types of End-Uses

Biofuels are used in the domestic sector as well as in institutions and small-scale industries. Woodfuel is used more than any other biofuel in all of these sectors (Leach and Mearns, 1988). The bulk of woodfuel is used in the domestic sector for cooking, space heating, ironing, and lighting. For example, a study undertaken in Sierra Leone by Kamara (1986) indicates that cooking accounted for 86% of the end-uses of fuelwood in the capital city of Freetown and 88% in rural Rwanda.

Space heating and lighting are in most cases external benefits associated with the use of fuelwood for cooking (Kgathi *et al.*, 1994). However, space heating as an end-use is common in countries that experience cold winters, such as South Africa, Lesotho, and Botswana. A study in Botswana, where average daily temperatures can sometimes drop to 5^{0}C, revealed that all socioeconomic groups used fuelwood for space heating in winter, and in areas where fuelwood was scarce, the rich peasants tended to use more fuelwood than the poor peasants (ERL, 1985). In a study on fuelwood consumption in Kenya, Hosier (1985) expected households in areas of high altitude to experience higher consumption levels of fuelwood. However, there was no evidence that ambient temperature variations had any effect on fuelwood consumption. It was concluded that "either the use of space heating does not vary throughout rural Kenya or else the relationship is obscured by other factors" (Hosier, 1985).

Lighting is not a popular end-use of fuelwood in Africa. It is most often associated with poor households that lack the means to purchase more advanced lighting fuels such as paraffin and candles. Studies in rural areas of Botswana indicate that only about 2% of the households use fuelwood for lighting (Kgathi, 1992; Kgathi *et al.*, 1994; Zhou, 1994).

In small-scale industries, the end-uses of woodfuel include tobacco curing, bread baking, fish smoking, tea drying and brick burning. These end-uses vary from country to country, depending on forest and energy resource endowments. For instance, tobacco curing is a common woodfuel end-use in tobacco-producing countries such as Malawi, Uganda, and Zimbabwe. On the other hand, tea drying is a common end-use in Tanzania (Kaale, 1990). In public institutions, woodfuel is mainly used for the preparation of meals for students in schools, as well as for patients in clinics and hospitals, particularly those that are located in rural areas. A recent study revealed that in the Botswana villages of Molepolole and Mmankgodi, fuelwood was used by most schools and health facilities interviewed, with schools the predominant users of this resource (Kgathi and Mlotshwa, 1994).

Crop residues and cow dung are also used in households, small-scale industries, and public institutions. In most cases, they are used only as supplementary fuels in the domestic sector (Kgathi, 1992), but in areas where there is an acute of scarcity of woodfuel, they may be used extensively (Barnard, 1990). The end-uses of crop residues and cow dung include household water heating, cooking, beer brewing, and brick making. In some cases, crop and animal residues are preferred for special end-uses, such as pottery, which requires low heat for a long time.

3.2. Technologies Used

A range of technologies is used to burn biomass in Africa (Agarwal, 1986). The type of technologies used depends on a variety of factors, such as the sector in which the biofuel is used, the type of end-use, and the level of income of the user. Most of the technologies for burning biomass are inefficient at converting wood into energy. Assuming that biomass is used for cooking, efficiency can be defined as the amount of energy transferred into the cooking pot divided by the amount of energy expended (Agarwal, 1986). In many parts of Africa, households burn biomass on open fires. The proportion of households using open fires is estimated to be more than 90% in most African countries. The fire is usually made between three stones or under an iron tripod. Open fires are widely used in Africa because they can be made easily and without cost. However, like most technologies used to burn biomass, open fires are criticized for low efficiency at converting wood into energy—5% to 10% (Earthscan, 1984).

It is important to note that open fires are used not only in households but also in small-scale industries and public institutions. Kgathi and Mlotshwa (1994) revealed that in all primary schools that used fuelwood in the villages of Molepolole and Khakhea, Botswana, meals were cooked in large, three-legged pots over open fires.

In some African countries, such as Tanzania, Zambia, and Senegal, traditional stoves are also used to burn biomass. These stoves are normally made of local materials, such as scrap metal and mud. Although they are normally used to burn fuelwood, they were originally designed to burn charcoal and tend to burn with higher efficiency than the traditional open fires.

Improved stoves have been introduced in many African countries because they are more efficient and, thus, conserve fuelwood (Earthscan, 1984). Laboratory tests indicate that improved stoves are not just thermodynamically efficient; they also save large amounts of fuelwood in everyday use. Comprehensive methods for assessing how much fuel is saved have been developed in recent years. It is estimated that these stoves save 16% to 63% of wood energy (Leach and Mearns, 1988). However research has shown that saving fuel is not the only factor stove users take into consideration. Other important considerations are reduced smoke, compatibility with present cooking patterns, reduced burns, and time savings. Unfortunately, improved stoves have not been widely and intensively disseminated in some countries. The only successful case is the Kenyan JIKO (a charcoal stove). In Kenya alone, 700 000 stoves have been sold. Similar stoves have been distributed in other eastern African countries, although not as many as the JIKO. In southern Africa, improved stoves are not widely used. The reason is that these stoves fail to provide other features (for example, the space heating and space for social gathering open fires can provide), and in some cases the stove designs were poor (Kgathi, 1992). It is clearly important to understand the needs of potential users before introducing innovations such as improved stoves.

Although improved stoves save wood energy, they may only marginally alleviate the problem of deforestation because this process is caused primarily by the clearing of land for agricultural purposes and by the exploitation of timber for commercial purposes (Krugman, 1987). However, charcoal production may lead to localized deforestation because it is more destructive to forest systems than is intermittent fuelwood collection (Krugman, 1987).

4. Magnitude of biomass consumption

Data on biomass consumption are scarce in most African countries, because, compared to modern fuels, biomass fuel is a very low priority in national development planning (Hall, 1994). Biomass is a noncommercial fuel and collecting data on these fuel types is quite difficult. In addition, the methods used to estimate the rate of biomass consumption are sometimes crude, so available data are not always accurate. In this section, an attempt has been made to examine the rates of biomass consumption in African households, small-scale industries, and public institutions.

4.1. Household Consumption

Per capita biomass consumption data has a very high inter- and intracountry variation. For instance, a survey conducted in fifteen villages of Tanzania revealed that fuelwood consumption varied from 0.45 to 1.6 Mg per capita annually (Table III). According to Kaale (1990), the average per capita fuelwood consumption in sub-Saharan African countries was estimated to be 1m^3 per year, or 0.9 Mg per capita annually in the 1980s (assuming 1.1 m^3 of fuelwood = 1 Mg fuelwood (Kaale, 1990)). However, this estimate seems high, if it is compared with the current per capita fuelwood consumption figures of most sub-Saharan African countries.

Fuelwood consumption is a dynamic rather than a static variable, because it tends to decrease as scarcity increases, as the next section on the determinants of fuelwood consumption will show. This relationship implies that the fuelwood consumption of a particular place will vary over time, depending on the levels of fuelwood scarcity.

In terms of the magnitude of consumption, fuelwood is the most important biofuel, followed by charcoal, dung, and crop residues (Hall, 1994; Leach and Mearns, 1985). In developing countries as a whole, it is estimated that woodfuel accounts for 83% of the total consumption of biofuels, followed by cow dung (15%) and crop residues (2%) (Arntzen and Kgathi, 1984).

TABLE III

Per capita bioenergy consumption in different African countries (tonnes per capita per year)

Country	Charcoal	Fuelwood	Bioenergy Residues	Dung	Source
Botswana (rural)	–	0.51	–	–	ERL (1985)
Botswana	–	0.53	–	–	Zhou (1994)
Botswana (Molepolole)	–	0.34	–	–	Kgathi and Mlotshwa (1994)
Rwanda	0.22*	0.33**	0.19**	–	Karenzi (1994)
Sudan (Mogadishu)	0.15	–	–	–	Bird and Sheppard (1989)
Swaziland	–	0.495	0.033	–	Eberhard (1984)
South Africa (KwaZulu)	–	0.62	–	0.012	Eberhard (1984)
South Africa (six rural villages)	–	–	–	0.05–0.23	Van Horen and Eberhard (1995)
Tanzania (15 villages)		0.45–1.6	–	–	Kaale (1990)
Zambia (rural)	0.02	1.2	–	–	Government of Zambia (1992)
Zambia (urban)	0.17	0.15	–	–	Government of Zambia (1992)

*charcoal consumption in urban Kigali, Rwanda.
** fuelwood consumption in rural Rwanda

Note: The fuelwood and dung figures for Botswana, South Africa, Swaziland, and Rwanda are in air dry weight, according to the sources of these figures. The sources for the Zambia and Tanzania figures did not specify whether the fuelwood is in oven dry, air dry, or green weight. In South Africa, dung was in sun dried weight (40% moisture content).

TABLE IV

Daily domestic consumption of wood and residues (wood and crop) per capita

Commune	Daily consumption per capita					Annual consumption per capita		
	Wood		Residues		Total	Total wood equivalent		
	kg	MJ	kg	MJ	MJ	MJ	Mg	m^3
Cyimbogo	0.94	14.10	0.64	8.96	23.06	8417	0.56	0.85
Mukingo	1.49	22.35	0.30	4.20	26.55	9691	0.65	0.98
Kiyombe	0.72	10.80	0.58	8.12	18.92	6906	0.46	0.69
Runyinya/ Nyaruheng/ Shyanda	1.10	16.50	0.70	9.80	26.30	9600	0.64	0.96
Ntyazo	0.30	4.50	0.75	10.50	15.00	5475	0.37	056
Mdasonwa	0.89	13.35	0.21	2.94	16.29	5946	0.40	0.60
Average	0.91	13.60	0.53	7.42	21.02	7672	0.51	0.77

MJ = Million joules
Note: kg are in dry weight. According to Karenzi, the moisture content of the fuel concerned was visually evaluated as either wet, half dry, or dry before measurements were done. Seventy-four percent used dry wood and 87% used dry residues.
Source: Karenzi (1994)

The contribution of different biofuels to the total energy consumption of biofuels, however, differs from country to country. In rural Rwanda, it was estimated that fuelwood accounted for 65% of household biomass energy consumption, whereas residues accounted for the remaining 35% (Table IV). Areas with a low consumption of fuelwood tended to use more residues than areas with a high consumption of fuelwood. For instance, in the commune of Ntyazo, daily per capita consumption of fuelwood was 0.30 kg compared with the commune of Mukingo, which consumed 1.5 kg. In the commune of Ntyazo, consumption was 0.75 kg residues per person per day, compared to Mukingo, where fuelwood was 0.30 kg. The annual per capita biomass consumption of the six communes in Rwanda was estimated to be 0.51 Mg of wood equivalent.

As previously stated, dung and crop residues are in most cases supplementary fuels, used when fuelwood is scarce (Barnard, 1990). Table III indicates that dung has been a common fuel in some parts of South Africa, with consumption estimates ranging from 0.05 to 0.23 Mg per capita per year in the six rural villages. These estimates are significantly lower than the average consumption of 0.4 Mg per capita in the homelands of South Africa (Van Horen and Eberhard, 1995).

Methods used to estimate household biomass consumption have differing levels of accuracy. Using the long-term monitoring methods of measuring biomass consumption, Karenzi (1994) estimated the daily per capita consumption of fuelwood in rural Rwanda to be 0.91 kg, or 0.33 Mg per year. This is lower than the estimate based on the user-recall method, estimated at 1.3 kg per day per person, or 0.4 Mg per capita per year. In addition, the annual per capita consumption of residues was estimated to be 0.19 Mg in the long-term monitoring method, compared with 0.5 Mg in the user-recall method.

In Botswana, a comparison of the results of the one-day measurement and the repetitive measurements revealed a variation in the rates of fuelwood consumption. The annual per capita fuelwood consumption in the village of Mmankgodi was estimated to be 0.39 Mg by the repetitive method compared with an estimate of 0.5 Mg obtained from the one-day measurement method.

The repetitive method is considered to be more reliable than the user-recall and the one-day measurement methods. The user-recall method relies on the user's memory, and the Rwanda study by Karenzi (1994) has shown that when this method is used, households have a tendency to overestimate biomass consumption. The one-day measurement method also tends to produce overestimates of biomass consumption, as shown by the Botswana study (Kgathi *et al.*, 1994), mainly because researchers tend to probe less when using this method. As a result, households that are temporarily using other sources of energy are often left out of the survey and treated as a missing value in the calculations. This method tends to yield a higher mean per capita estimate (Kgathi *et al.*, 1994).

4.2. Small-Scale Industries, Commercial Enterprises, and Public Institutions

The amount of biomass used in the nonhousehold sector varies according to the type of enterprise, the institution and production process undertaken, the scale of operation, and the efficiency of the equipment used (Kgathi and Mlotshwa, 1994). As a small-scale industry, the tobacco industry is known to be one of the major consumers of fuelwood in Africa. This is exacerbated by the industry's tendency to use inefficient technologies that over-use fuelwood. In Tanzania, it is estimated that curing one hectare of tobacco will require that about one hectare of woodland be cut as fuelwood (MnZava, 1981). Tobacco curing accounts for 23% of the total fuelwood consumption in Malawi. In Tanzania, tea drying is estimated to consume 43 600 m^3 of fuelwood per year (MnZava, 1981). To reduce fuelwood consumption, some farmers are resorting to more efficient technologies that use coal. In Zimbabwe, for example, commercial farmers are moving away from the use of traditional biomass-based tobacco barns to more efficient coal barns.

The amount of biomass used in other small-scale industries is not available, but there is additional data on the amount of end-products produced per 100 kg of fuelwood (Table V). This information may be used to estimate the amount of biomass used in a particular country, if data on the national production of commodities are available and assuming a static technology status. Similarities in beer brewing figures for Burkina Faso and Zimbabwe suggest that product/fuelwood might be a good unit of analysis for determining the magnitude of biomass consumption in other small-scale industries.

Data are also scarce regarding the extent to which biofuels are used in public institutions and commercial enterprises. A few available studies suggest that they use large amounts of biofuels. For example, in urban Tanzania substantial amounts of woodfuels were used in the preparation of meals in restaurants, hotels, and hospitals. In Dar es Salaam, the total

TABLE V

Quantitative comparisons of industrial products produced by 100 kg of fuelwood using traditional methods

Country	Amount of End-Products
Malawi	cures 6 kg of tobacco
Tanzania	cures 38 kg of green leaves
Ivory Coast	smokes 66 kg of fresh fish
Burkina Faso	brews 100 l of traditional beer
Zimbabwe	brews 123 l of traditional beer
Zimbabwe	makes 108 kilns of bricks

Source: Kgathi and Mlotshwa (1994)

consumption of charcoal was 148 800 GJ compared with 89 160 GJ of fuelwood. The values for the smaller city of Mbeya were 17 500 GJ for charcoal and 157 000 GJ for fuelwood (Luhanga and Bwakea, 1994). Another study, undertaken in the large village of Molepolole in Botswana, revealed that public institutions and restaurants consumed a total of 30 407 kg of fuelwood per month, which is equivalent to an annual value of 365 Mg. A large proportion of this amount (89%) was consumed by primary and secondary schools, and the rest (11%) was consumed by medical clinics and restaurants (Kgathi and Mlotshwa, 1994).

5. Determinants of biomass consumption

Information is generally lacking on the determinants of biofuel use in Africa, except for determinants of fuelwood consumption. Research in Africa indicates that domestic fuelwood consumption is determined by a number of factors such as fuelwood scarcity, household income, household size, and climatic conditions (Hosier, 1985; Kgathi, 1992; Kgathi *et al.*, 1994, Kamara, 1986). This section focuses on fuelwood and discusses determinants of fuelwood consumption.

5.1. Fuelwood Scarcity

Fuelwood consumption tends to decrease as its scarcity or the cost of its collection increases (Leach and Mearns, 1988). The price of fuelwood or the labor time expended collecting it is usually used as a proxy for fuelwood scarcity. A study undertaken in Botswana revealed an inverse relationship between fuelwood consumption and its price. For instance, using the one-day measurement method, the mean household fuelwood consumption in the large village of Mochudi was estimated at 8 kg per day and only 5.8 kg per day in the capital city of Gaborone. In Mochudi, where fuelwood consumption was higher, the price of a small bundle of purchased fuelwood was approximately one-half the price in Gaborone.

Another study, undertaken by Oki (1985) in the southeastern Botswana villages of Shoshong and Ditshegwane, revealed an inverse relationship between fuelwood consumption and the time it takes to collect it. According to this study, per capita fuelwood consumption was estimated to be 0.99 kg in Ditshegwane, and 0.60 kg in the large village of Shoshong. The distances traveled for fuelwood collection, which are a proxy for the labor time on fuelwood collection, were estimated at 7 km in Shoshong and 3.5 km in the smaller village of Ditshegwane. In Kenya, Hosier (1985) revealed that when the time of collecting one kg of fuelwood increased by one minute, fuelwood consumption decreased by 171 kg annually.

The inverse relationship between fuelwood consumption and its price or collection time suggests that households tend to use fuelwood more sparingly as the economic cost of collecting it increases (Kgathi *et al.,* 1994). Thus, when fuelwood is collected nearby, households tend to increase consumption to the maximum possible, because the labor time, and hence the opportunity cost, for its collection are low. This phenomenon occurs until the marginal utility is very low indeed. However, as the distances to the points of fuelwood collection increase, as does the opportunity cost of its collection, consumption tends to decrease (Leach and Mearns, 1988).

5.2. Income

The relationship between income and fuelwood consumption has been widely studied in Africa (Hosier, 1985; Kgathi, 1992; Kgathi *et al.,* 1994). Using data from the Household Income and Expenditure Survey: 1985/86, Kgathi (1992) found that income elasticities of the various income groups in rural Botswana were positive, implying that an increase in household incomes is associated with an increase in fuelwood consumption. These results were confirmed when another study showed that an increase in income of a household, as proxied by the number of cattle owned, was associated with an increase in fuelwood consumption in the village of Khakhea, Botswana (Kgathi *et al.,* 1994).

In urban areas, the income elasticity of fuelwood demand tends to be positive in low income groups. This is because households do not normally meet their energy requirements, and as their incomes increase, they tend to consume more fuelwood (Kgathi, 1992). Among the middle and high income groups, income elasticity tends to be negative, because these households tend to consume the more technologically advanced fuels as their incomes increase.

5.3. Household Size

Research conducted in a number of African countries such as Kenya, Sierra Leone, and Botswana reveals an inverse relationship between per capita fuelwood consumption and household size (Kgathi, 1992; Kgathi *et al.,* 1994; Kamara, 1986; and Hosier, 1985). This inverse relationship exists because large households tend to use fuelwood more efficiently than small households. Kamara (1986) referred to this relationship as "communal use efficiency" of energy. Communal use efficiency was not only associated with fuelwood in Sierra Leone but also with other energy sources, such as kerosene and electricity.

This fuelwood use-efficiency relationship could also be viewed as a mathematical, economy-of-scale issue. Household energy consumption is not particularly sensitive to increases in household sizes, because if one or two members join a household, cooking and space heating do not necessarily increase.

6. Interfuel substitution

Data on the patterns of fuel-switching in African households are limited, and this section is based on evidence from only a few African countries. A 1987 survey undertaken in Dar es Salaam, Tanzania, revealed that 38% of the households had switched from kerosene to charcoal over the previous two years (Leach and Mearns, 1988). In addition, 10% of the households had switched from charcoal to fuelwood, and 15% from fuelwood to charcoal. No households were reported to have switched from biofuels to commercial sources of energy. There was a net gain in the number of households that had switched down the energy ladder to biofuels. The main reason given for the switch away from traditional fuels was difficulty with supply. The reasons given for switching to the new fuels were that they were cheaper (25%), convenient (17%), easy to use/obtain (30%) and cheaper/ easy to use (14%) (Leach and Mearns, 1988).

Studies in Botswana revealed that most households had switched from fuelwood to other commercial sources of energy, such as kerosene and liquefied petroleum gas (LPG), during the five-year period before the study (Kgathi, 1992; Kgathi *et al.*, 1994; ESMAP, 1991). Kgathi (1992) found that out of the households that had switched fuels in the capital city of Gaborone, 38% had switched from fuelwood to kerosene and 25% from fuelwood to LPG. In the large village of Mochudi, 60% of the households had switched from fuelwood to kerosene and 32% from fuelwood to LPG. Some of the households had partially switched from fuelwood to wastes. They did not switch from fuelwood to cow dung and crop residues, as is common in most rural areas of Africa. This is probably because these energy sources were not available nearby. Households switched from fuelwood to dung in Good Hope, Botswana, where commercial agriculture has depleted the surrounding woodlands. It was estimated that 82% of the households used cow dung as fuel (ERL, 1985). Up to 69% of households switched away from fuelwood in Gaborone and Mochudi primarily because fuelwood was scarce. The second major reason for switching was convenience (for example, to LPG) (19%).

The main factors behind fuel switching are the availability of the fuels and sufficient income to purchase the devices for using them (Leach and Mearns, 1988). Case studies of Tanzania and Botswana reveal a relationship between household fuel switching and these two main factors. If, for instance, real wages are declining, households will find it difficult to purchase expensive fuels and the devices for using them, and they may switch to the cheaper fuels. The probability of fuel switching is even higher if supply of the fuels being used is insecure.

7. Implications for GHG emissions and mitigation strategies

The type of biofuel used, its end-uses, the efficiency of the technologies employed to produce and use the biofuel, and the quantities of fuel consumed have a direct bearing on the scale of GHG emissions produced (Earthscan, 1984). When the proportion and scale of use of the different biofuels are known it is possible to calculate fairly accurate GHG emissions. However, data on fuelwood are crude and mostly pertain to household use. In the household, especially households in rural areas, the fact that cooking and space and water heating tend to be concurrent may result in double counting of emissions, if the end-uses are taken as separate consumers of fuel. It is also important to know the proportions of households using stoves and the efficiency of the various stoves so estimates of fuel savings can be derived and avoided GHG emissions calculated.

8. Conclusions

Data on the patterns of biomass consumption in Africa are insufficient to derive accurate greenhouse gas (GHG) emission estimates. Because some of the estimates on biomass consumption may be crude, the tendency to use any available data can lead to inappropriate policy decisions regarding mitigation options (Hall, 1994).

Cooking is the most important end-use of biofuels in Africa, implying that it is responsible for most of the GHG emissions. Mitigation strategies should, therefore, be directed at improving cooking techniques and devices with the aim of lowering GHG emissions. For example, the use of the improved cooking stoves should be encouraged in households, small-scale industries, and public institutions because they save large amounts of biofuels, and thereby reduce the emission of GHGs such as CO, N_2O, and NO_X. It is also true, however, that compared with open fires, the use of more efficient stoves has been associated with an increase in the emission of CH_4 (Hall and Rosillo-Calle, 1991).

This paper has also shown that biomass consumption is a dynamic rather than a static variable, implying that the consumption levels in a particular place will vary over time, depending on the levels of scarcity (Earthscan, 1984). Consequently, studies that base biomass consumption projections only on population, keeping per capita consumption constant, have serious methodological flaws. It is likely that such studies will also erroneously project GHG emissions. The only feasible way of dealing with projected GHG emissions is to construct socioeconomic scenarios. Usually, three scenarios are assumed—the business-as-usual, the worst-case (assuming largest biofuel consumption level) and the best-case scenarios (assuming the least consumption level). The mitigation options associated with each of the scenarios can then be evaluated.

Finally, in many African countries interfuel substitution is common in the household sector. Households have switched to biofuels and to commercial fuels. One of the major factors behind energy transitions in Africa is the level of income. In countries where incomes have been declining over time, households have switched from commercial sources of energy to biofuels. The trends in interfuel substitution have implications for GHG emissions and mitigation strategies in developing countries.

References

Abel, N.O.J., *et al.*: 1988, *Amelioration of the Soils by Trees: Guidelines for Training in Rapid Appraisal for Agroforestry Research and Extension*, Harare Forestry Commission and London Commonwealth Science Council.

Agarwal, B.: 1986, *The Woodfuel Crisis in the Third World*, London: Zed Books, Ltd.

Ardayfio-Schandorf, E.: 1993, Fuelwood Energy Crisis in Sub-Saharan Africa, Conference on Sustainable Environmental and Resource Management Futures, Accra, Ghana.

Arntzen, J.W. and Kgathi, D.L.: 1984, Some of the determinants of the consumption of firewood energy in developing countries: the case of rural Botswana, *Pula: Journal of African Studies*, **4**, no.1, 24–34.

Barnard, G. W.: 1990, Use of agricultural residues as fuel, *Bioenergy and the Environment*, Pasztor, J. and Kristoferson (eds.), Oxford, U.K.: Westview Press.

Bhatia, R.K.: 1985, Traditional energy sources data, in Integrated Energy Planning: A Manual, Kuala Lumpur, Asian Development Centre.

Bird, N.M. and Shepperd, G.: 1984, Charcoal in Somalia: A Woodfuel Inventory in the Bay Region of Somalia, London, Overseas Development Assistance.

Brokensha, D. and Castro, A.P.: 1993, Methods of Fact Finding, in *Woodfuel Surveys*, Rome: Food and Agriculture Organization (FAO).

Davidson, O.R.: 1985, *Energy Use Patterns Sierra Leone*, Ottawa: International Development Research Centre (IDRC), Ottawa, Canada.

Digernes, T.H.: 1977, Wood for Fuel: Energy Crisis Implying Desertification, the Case of Bara, the Sudan, Ph.D. Thesis, University of Bergen, Norway.

Earthscan: 1984, *Fuelwood: Which Way Out*, London: Earthscan.

Eberhard, A.A.: 1984, Energy Use, Problems, and Supply Strategies for Underdeveloped Areas in Southern Africa Energy for Underdeveloped Areas, Cape Town, Energy Research Institute.

Eberhard, A.: 1986, *Energy Consumption Patterns in Underdeveloped Areas in South Africa*, Cape Town: Energy Research Institute, University of Capetown.

ERL (Energy Resources Limited): 1985, A Study of Energy Utilization and Requirements in the Rural Sector of Botswana, London: Energy Resources Limited.

Energy Sector Management Assistance Program (ESMAP): 1991, Botswana Urban Household Energy Strategy, Washington D.C.: Industry and Energy Department, The World Bank.

FAO: 1983, *Woodfuel Surveys*, Rome: FAO.

Gay, J. and Zietlow, K.: 1985, Botswana Urban Domestic Energy Use and Attitude Survey, Ministry of Mineral Resources and Water Affairs, Gaborone.

Government of Zambia: 1992, Energy Statistics Bulletin, 1974–1990, Department of Energy, Ministry of Energy and Water Development, Lusaka.

Hall, D. and Rosillo-Calle, F.: 1991, Why Biomass Matters: Energy and the Environmental, *Biomass Users Network, NetWork News* **5**, no.4.

Hall, D.O.: 1994, Introduction, summary, and conclusions, *Biomass Energy and Coal in Africa*, Hall, D.O., and Mao, Y.S., (eds.), London: Zed Books, Ltd.

Hosier, R.: 1985, *Energy Use in Kenya: Household Demand and Rural Transformation*, Stockholm: The Beijer Institute, Stockholm, Sweden.

Hosier, R. and Dowd, J.: 1988, Household energy use in Zimbabwe, in *Energy for Rural Development*, Hosier, R.H. (ed.), Beijer Institute, Stockholm, Sweden.

Kaale, B.K.: 1986, Biomass Research Assessment for Energy Production in Tanzania, in Assessing Biomass Energy Resources in Developing Countries. London: Commonwealth Science Council.

Kaale, B.K.: 1990, Traditional fuels, in *Bioenergy and the Environmental*, Pasztor, J. and Kristoferson, L. A. (eds.), Oxford, U.K.: Westview Press.

Kamara, J. N.: 1986, Firewood Energy in Sierra Leone—Production, Marketing and Household Pattern, Published Ph.D. Thesis, Goettingen University, Hamburg: Verlag Weltachiv GmbH.

Karenzi, P.C.: 1994, Biomass in Rwanda, in *Biomass Energy and Coal in Africa*, Hall, D.O. and Mao, Y.S. (eds.), London: Zed Books, Ltd.

Kgathi, D.L.: 1989, A Critical Review of Socio-Economic Surveys on Fuelwood in Botswana, a paper prepared for the Workshop on Energy Policy and Planning in Eastern and Southern Africa, March 11 to 17, 1989.

Kgathi, D.L.: 1990, A critical review of fuelwood surveys in Botswana in *African Energy Issues in Planning and Practice*, African Energy Policy Research Network (AFREPREN), London: Zed Books, Ltd.

Kgathi, D.L.: 1992, Household Response to Fuelwood Scarcity: Implications for Energy Policy, Ph.D. Thesis, University of East Anglia.

Kgathi, D.L. and Mlotshwa, C.V.: 1994, *Utilization of Fuelwood in Botswana: Implications for Energy Policy*, Nairobi: AFREPREN.

Kgathi, D.L., Sekhwela, M.B.M., Tietema, T., and Mpotokwane, M.A.: 1994, Biomass in Botswana, in *Biomass Energy and Coal In Africa*, Hall, D.O. and Mao, Y.S. (eds.), London: Zed Books, Ltd.

Krugman, H.: 1987, *Review of Issues and Research Relating to Improved Cookstoves*, Ottawa: IDRC.

Leach G. and Mearns, R.: 1988, *Beyond the Fuelwood Crisis*, London: Earthscan Publications, Ltd.

Luhanga, M.L. and Bwakea, T.: 1994, *Energy Consumption in the Urban Commercial Sector in Tanzania*, Stockholm: Stockholm Environment Institute.

MnZava, E.M.: 1981, Wood energy, *UNASYLA* **33:** 131.

Mwandosya, M.J.: 1986, Bioenergy research assessment, in *Assessing Biomass Energy Resources in Developing Countries*, London: Commonwealth Science Council.

Oki, J.A.: 1985, Domestic Energy Use in Two Villages, Addendum to Botswana Village Energy Survey, Gaborone: Ministry of Mineral Resources and Water Affairs.

Tietema, T., Kgathi, D.L., and Merkesdal, E.: 1988, Wood Production and Consumption in Dukwe, a feasibility study for a woodland management and plantation scheme, Gaborone: National Institute of Development Research.

Tomkins, R.: 1986, A review of traditional fuel survey methods, in *Assessing Biomass Energy Resources in Developing Countries*, London: Commonwealth Science Council.

UNEP, OECD, IEA, IPCC (United Nations Environment Programme, Organisation for Economic Co-operation and Development, International Energy Agency, Intergovernmental Panel on Climate Change): 1995, *IPCC Guidelines for National Greenhouse Gas Inventories*, IPCC, Bracknell, 3 Volumes.

Van Horen, C. and Eberhard, A.: 1995, Energy, Environment and the Rural Poor in South Africa, Energy for Development Research Centre. Rondebscoch, South Africa.

Zhou, P.: 1994, Domestic Energy Consumption and Its Impact on the Environment in Botswana, African Development Bank.

NOTE

CARBON-TO-NITROGEN RATIOS IN AGRICULTURAL RESIDUES

J.O. ILUKOR and S.O. OLUKA

Department of Physics, Makerere University, P.O. Box 7062, Kampala, Uganda

Abstract. Agronomic crop residues produce greenhouse gas emissions. Crops that produce residues at harvest and during processing may vary from country to country. These residues, which can be in the form of peels, husks, stalks, or straw, are generally considered to be waste products. The carbon (C) and nitrogen (N) content of 19 different agronomic and grass crops common in Uganda were determined using standard laboratory methods. The C and N content of the samples were calculated from two separate equations containing a moisture correction factor. The crop residue C/N ratios were similar to UNEP/OECD/IEA/IPCC values.

1. Introduction

The *IPCC Guidelines for National Greenhouse Gas Inventories* (UNEP *et al.*, 1995) provide a default methodology for quantifying sources and sinks of key anthropogenic greenhouse gases (GHGs). Because this methodology is generic, default values for specific sectors (for example, energy or land use) can lead to significant errors in GHG emission estimates. Emissions from the land-use sector (agriculture and forestry) are particularly difficult to estimate because of the complex biogeochemistry of ecosystems (Dixon *et al.*, 1994). Combustion of biomass and biomass chemistry are two factors that are especially difficult to characterize (Levine, 1991).

Biomass varies in its carbon (C) and nitrogen (N) content, depending on the plant species, edaphic factors, climate, and cultivation practices in any given region (Seiler and Crutzen, 1980). This variation influences the net heating values of biomass combustion as well as the characteristics and intensity of GHG emissions (Crutzen and Andreae, 1990). The purpose of this study was to evaluate C and N content of plant tissues common in Uganda and compare values to UNEP/OECD/IEA/IPCC methodology.

2. Materials and methods

A variety of agronomic crop and grassland ecosystem biomass commonly combusted by wildfire was analyzed (Tables I and II). Samples were collected using a stratified sampling system for representative ecosystems. Nineteen different plant types common in Uganda were sampled.

The samples from each plant residue were powdered. The C content of each powdered sample was determined by oxidizing the sample in a digestion tube at 120 degrees celsius by adding both potassium dichromate solution and concentrated sulfuric acid (Crutzen and

Environmental Monitoring and Assessment **38**: 271-275, 1995.

TABLE I

Biofuel parameters for grass and crop residues found in Uganda ecosystems

Plant tissue	C Content		N Content		C/N Ratio		Energy Output Gj/Mg		Residue/Product Ratio	
	$\bar{x}$	sd	$\bar{x}$	sd	$\bar{x}$	sd	$\bar{x}$	sd	$\bar{x}$	sd
Sugar cane (husks) (*Saccharum L.*)	0.347	0.07	0.0035	0.003	99.1	10.2	8.8	1.1	0.3	0.1
Coffee (husks) (*Coffea L.*)	0.381	0.08	0.0175	0.002	21.8	2.2	8.9	2.3	0.8	0.2
Cotton (stalks) (*Gossypium L.*)	0.342	0.06	0.0192	0.002	17.8	1.9	9.1	1.9	12.5	1.2
Grass (mixed) (Gramineae)	0.305	0.08	0.0183	0.007	16.7	3.1	7.9	1.9		
Papyrus (*Cyperus*)	0.314	0.03	0.0099	0.004	31.7	2.8	7.3	1.3		
Elephant grass (*Typha*)	0.269	0.06	0.0140	0.005	19.2	1.9	8.0	1.6		

Andreae, 1990). The C content (C%) of the sample was then calculated using the following equation:

$$C\% = \{[3(b - s)]/b/W\} \times M \times 100$$

where

b	=	ml of $FeSO_4$ used for the blank titration.
s	=	ml of $FeSO_4$ used for the sample titration.
3	=	equivalent weight of C.
W	=	weight of sample.
M	=	moisture correction factor.

The N content of biomass was similarly evaluated. Compounds containing N in the biomass sample are reduced to NH_3 compounds by Kjeldhal standard digestion procedure (Vitousek, 1982). The N content of the sample was calculated using the following equation:

$$N\% = [(b - s)\ m/W] \times 14 \times (50/V) \times M \times 100$$

where

b	=	ml of KH(103)2 used for blank titration.
S	=	ml of KH (103)2 used for sample titration.
M	=	molarity of the KH(103)2.
14	=	atomic weight of N.
50	=	ml volume of digestion tube.
V	=	ml volume of the digest.

TABLE II

Biofuel parameters for agronomic crop residues in Uganda

Plant tissue	C Content		N Content		C/N Ratio		N/C Ratio		Energy Output GJ/Mg	
	x̄	sd	x̄	sd	x̄	sd	x̄	sd	x̄	sd
Bananas (*Musa* L.)	0.312	0.08	0.0090	0.003	34.7	9.3	0.0288	0.01	7.6	1.3
Millet (*Panicum* L.)	0.326	0.04	0.0160	0.003	20.4	2.6	0.0491	0.01	8.4	1.1
Maize (*Zea* L.)	0.329	0.06	0.0053	0.002	62.1	12	0.0161	0.01	8.9	1.2
Sorghum (*Sorghum* L.)	0.284	0.07	0.0110	0.003	25.8	6.3	0.0387	0.03	8.3	1.1
Rice (*Oryza* L.)	0.248	0.07	0.0120	0.003	20.7	4.1	0.0484	0.01	8.1	1.2
Wheat (*Triticum* L.)	0.314	0.02	0.0140	0.002	22.4	3.1	0.0446	0.01	7.2	1.3
Potatoes (*Solanum L.*)	0.365	0.08	0.0123	0.004	29.7	6.4	0.0337	0.01	7.9	1.4
Cassava (*Manihot*)	0.332	0.09	0.0230	0.004	14.4	2.5	0.0693	0.12	8.3	2.1
Beans (*Vicia*)	0.297	0.06	0.0070	0.002	42.4	8.4	0.0236	0.01	7.7	1.6
Peas (*Pisum*)	0.297	0.05	0.0330	0.003	9.0	1.0	0.1110	0.01	7.6	1.3
Peanuts (*Arachis*)	0.296	0.08	0.0140	0.003	21.1	5.2	0.0473	0.12	8.3	1.2
Soybeans (*Glycine L.*)	0.335	0.04	0.0210	0.002	16.0	1.7	0.0627	0.01	7.8	1.7

W = weight of the sample.
M = moisture correction factor.

Dry matter weight, as well as the residue-to-crop-product ratio, was determined from measurements of dry biomass samples (von Malitz and Scholes, 1995). The C content in dry matter is higher than that of wet matter. For the agricultural crop products sampled, an average value of C content per unit of dry matter was determined to be 0.33. Therefore, to convert from dry matter to C content, a mean value of 0.35 was used for estimates in cases where crop-specific C content data were not available.

To determine the net caloric content of each of the agricultural crop residue, quantities of samples were completely burned in a combustion chamber. The net heating value (equal to the grass heating value less the heat carried away by the products of combustion) was estimated by the calorimetric method (Crutzen and Andreae, 1990).

The aboveground biomass was determined by specific field measurements of dry matter. This allowed the residue-to-crop-product ratio to be estimated for the agronomic crops. In Uganda, about 32% of the total crop residue surveyed is burned annually. This proportion compares with the average value of 25% for developing countries and 10% for developed countries (UNEP *et al.*, 1995).

The main source of sampling error is the intrinsic uncertainty of the water content of the crop residue. There are also errors in the determination of what exactly constitutes residue compared to the crop product when there are harvest losses. Subsamples enabled the standard deviation to be calculated to two significant digits for each crop.

3. Results

The C content of grass tissue ranged from 0.25 to 0.38 (Tables I and II). Among agronomic crops, C content was very homogeneous. The N content of the plant tissue ranged widely. As a result, C/N ratios varied from 99 for sugar cane to 17 for mixed grass. The energy content was similar for all plant tissues sampled.

The crop residue ratio and N/C values from the current study and UNEP *et al.* (1995) N/C default factors are compared in Table III. Crop residue C values from this study are similar to UNEP/OECD/IEA/IPCC values. In contrast, the UNEP/OECD/IEA/IPCC values for N/C differ considerably from the current study.

4. Discussion and conclusion

The C/N ratios and other biomass parameters were evaluated in this study to characterize agricultural crop and grass residues in Uganda. Country-specific C- and N-content data are needed to estimate GHG emissions from savanna burning and agricultural crop residues in Uganda. This information on ecosystem C and N biogeochemistry will help improve estimates of GHG emissions from land-use change in developing countries (Dixon *et al.*, 1994). Future country GHG inventories should include efforts to generate specific country emission factors to validate or calibrate UNEP *et al.* (1995) default values.

Acknowledgments

We greatly appreciate the support from the UNEP GHG Emission Inventory Study, which was coordinated by the Uganda Department of Meteorology, Ministry of Natural Resources, Uganda. Our thanks go to Makerere University for the laboratory facilities that were used for sample analysis.

TABLE III

Crop residue ratio and N/C values from current study and UNEP/IPCC/OECD/IEA

Crop	Residue Ratio	Uganda Residue Ratio	IPCC N/C Ratio	Uganda N/C Ratio
Wheat (*Triticum* L.)	1.3	1.2	0.012	0.0446
Maize (*Zea* L.)	1	1	0.02	0.0161
Rice (*Oryza* L.)	1.4	1.4	0.014	0.0484
Millet (*Panicum* L.)	1.4	1.1	0.016	0.0491
Sorghum (*Sorghum* L.)	1.4	1.2	0.02	0.0387
Soybeans (*Glycine* L.)	2.1	0.7	0.05	0.0627
Peanuts (*Arachis*)	1	1	N/A	0.0473
Potatoes (*Solanum* L.)	0.4	0.4	N/A	0.0507

References

Crutzen, P.J., and Andreae, M.O.: 1990, Biomass burning in the tropics: Impact on atmospheric chemistry and biogeochemical cycles, *Science* **250,** 1669–1678.

Dixon, R.K., Brown, S., Houghton, R.A., Solomon, A.M., Trexler, M.C., and Wisniewski, J.: 1994, Carbon pools and flux of global forest ecosystems, *Science* **263,** 185–190.

Levine, J.S. (ed.): 1991, *Global Biomass Burning: Atmospheric, Climatic and Biospheric Implications,* Cambridge, Massachusetts: MIT Press.

Seiler, W., and Crutzen, P.J.: 1980, Estimates of gross and net fluxes of carbon between the biosphere and the atmosphere from biomass burning, *Climatic Change* **2,** 207-247.

UNEP, OECD, IEA, IPCC (United Nations Environment Programme, Organisation for Economic Co-operation and Development, International Energy Agency, Intergovernmental Panel on Climate Change): 1995, *IPCC Guidelines for National Greenhouse Gas Inventories,* IPCC, Bracknell, 3 Volumes.

Vitousek, P.M., Gogz, J.R., Grier, C.C., Melillo, J.M., Reiner, W.A.: 1982, A comparative anlaysis of potential nitrification and nitrate mobility in forest ecosystems, *Ecological Monograph* **52,** 155–177.

von Malitz, G.P., and Scholes, R.J.: 1995, The burning of fuelwood in South Africa: When is it sustainable? *Environmental Monitoring and Assessment.* in press.

PART IV

GREENHOUSE GAS EMISSION INVENTORIES: NATIONAL ASSESSMENTS

FUEL CHARACTERISTICS AND EMISSIONS FROM BIOMASS BURNING AND LAND-USE CHANGE IN NIGERIA

A.O. ISICHEI[1], J.I.MUOGHALU[1], F.A. AKEREDOLU[2], and O.A. AFOLABI[3]

Departments of [1]Botany and [2]Chemical Engineering, Obafemi Awolowo University, Ile-Ife, [3]Federal Environmental Protection Agency, Abuja, Nigeria

Abstract. Nigeria is one of the 13 low-latitude countries that have significant biomass burning activities. Biomass burning occurs in moist savanna, dry forests, and forest plantations. Fires in the forest zone are associated with slash-and-burn agriculture; the areal extent of burning is estimated to be 80% of the natural savanna. In forest plantations, close to 100% of litter is burned. Current estimates of emissions from land-use change are based on a 1976 national study and extrapolations from it. The following non-carbon dioxide (CO_2) trace gas emissions were calculated from savanna burning: methane (CH_4), 145 gigagrams (Gg); carbon monoxide (CO), 3831 Gg; nitrous oxide (N_2O), 2 Gg; and nitrogen oxides (NO_X), 49 Gg. Deforestation rates in forests and woodlands are 300×10^3 ha (kilohectare, or kha) and $200 \times$ kha per year, respectively. Trace gas emissions from deforestation were estimated to be 300 Gg CH_4, 2.4 Gg N_2O, and 24 Gg NO_X. CO_2 emissions from burning, decay of biomass, and long-term emissions from soil totaled 125 561 Gg. These estimates should be viewed as preliminary, because greenhouse gas emission inventories from burning, deforestation, and land-use change require two components: fuel load and emission factors. Fuel load is dependent on the areal extent of various land uses, and the biomass stocking and some of these data in Nigeria are highly uncertain.

1. Introduction

The ecoclimatic zones in Nigeria range from the very humid freshwater and mangrove swamps in the south to the semiarid Sahelian zone in the north (Keay, 1959). The most extensive vegetation zones are forests and savannas. Savanna, a seasonal tropical vegetation type in which there is a closed or nearly closed groundcover of grasses that have flat, cauline leaves and are at least 80 centimeters (cm) high, comprises close to 80% of Nigeria's area of 924 000 square kilometers (km^2). Trees and shrubs are most often present in savanna, and they burn periodically (Sanford and Isichei, 1986).

Savannas can be divided into two broad groups: moist (humid) and dry (subhumid). Most burning takes place in the moist savannas, where the moisture-indexed length of the growing season is between 151 and 270 days and the rainfall/evapotranspiration ratio is between 0.4 and 1.0. This zone is described as subhumid (ILCA 1979). In terms of vegetation classification, the savanna area includes the derived (Transition), southern guinea, and northern guinea zones, or the guinea-congolian mosaic of lowland rainforest; *Isoberlinia* woodland and secondary grassland; and Sudanian woodland with abundant *Isoberlinia* (White, 1983; Keay, 1959).

Nigeria has a population of 88 million people and is one of the most densely populated countries in Africa. Most of the country is suitable for agriculture, and land use is largely driven by fiber, fuel, and food production. Fire is used extensively for bush clearing and

Environmental Monitoring and Assessment **38**: 279-289, 1995.

weed control in the widely practiced subsistence agriculture, and as a range management tool in livestock systems. Fire use in vegetation systems is human-related, but natural fires caused by lightning have been reported in savannas (Menaut *et al.*, 1985). Fires are so intimately associated with savannas that they are regarded as fire climax ecosystems. Fire is more essential to savanna maintenance than a humid climate. Fire frequency decreases with decreasing rainfall, from virtually annually in humid savannas to occasional and irregular fires in arid regions (once in 20 years) (Walker, 1985; Skarpe 1992). All plant species in the humid savanna communities are adapted to fires, without which the grass sward degenerates through the accumulation of senescent material and death of tillers. Savanna fires are generally restricted to the herbaceous layer, and crown fires are infrequent. In humid savannas the fuel load and intensity of the fires are such that the vegetation develops into a mosaic of open grassland with clumps of closed woodland and forest. Derived, or transition, savanna in West Africa converts back to forest if it remains unburned for some time.

Extensive wildfires are rare in the forest zone except in dry forests and plantations that suffer ground (litter) fires (Delmas *et al.*, 1991). Fires in the forest zone are associated with slash- and-burn agriculture, charcoal production, and conversion of forest to other permanent land uses, such as plantations of crops and exotic trees. One of the effects of forest fires is the transformation of dry forests into savanna. Once established, such savannas are maintained by fires (Swaine, 1992). Whether burned directly or converted, the loss of tropical forest cover has negative consequences for future global climate (Brown *et al.*, 1993).

Dry forest fires combust forest floors and litter and often smolder rather than burn openly. Small plants, including juvenile trees, suffer the greatest mortality, and large trees suffer bole damage and may die within 2 to 4 years after the fire. Opening the forest canopies allows the germination of certain herbaceous and perennial shrubs, especially plants in the family *Zingiberaceae*, and the shrub *Chromolaena odorata*, which is the best indicator of disturbed humid vegetation in West Africa (Keay, 1959).

Humid savanna fires mostly occur when vegetation is not completely dry. Delmas *et al.* (1995 a,b) report a grass moisture content of 7% to 8% in the dry season in the humid Lamto savanna in Côte d'Ivoire, as compared with 4.2% at Nylsvley in South Africa (Frost, 1985). Thus, only leaves, which are drier than stems, burn in early savanna fires. In dry forests where *Chromolaena* dominates, only the leaves burn. In humid savannas, up to 80% of the vegetation cover may burn, whereas in the Sudan zone only 25% to 50% (Delmas *et al.*, 1991) burns. An even lower proportion of litter burns in the Sahel.

The burning of savanna is linked to vegetation phenology. Savanna plant growth starts with the sprouting or regrowth (flushing) from rhizomes whose tops had been burned the previous season (Afolayan, 1978). Flushing can start as early as January (the middle of the dry season) and is a major reason why nomadic herdsmen burn savannas, the new flush being rich in nutrients (Collins, 1977). Germination of annual grasses starts with the rains, which in the Nigerian savanna begin between April and late May. Grasses attain their peak growth in October-November, after which they start senescing and drying up.

Burning can start in mid-December and by the end of January, most burning has taken place. The longer the burning is delayed into the dry season, the more intense the burn and the less the nutrient content of the grasses (due to translocation of some nutrients to belowground parts) (Isichei, 1983). Most West African savanna trees shed their leaves and flush new growth between February and May so that only a small proportion of savanna leaf litter is burned in annual fires (Collins, 1977).

Forest plantation fires (especially *Tectona grandis* and *Gmelina*) also occur in December-January. Plantation fires are more intense because the trees are completely deciduous and litter production is abundant (Akeredolu and Isichei, 1991). Leaf shedding occurs over a short time in December-January concomitant with occurrence of fires. Up to 100% of the litter is burned in some forest plantation fires.

Nigeria is reportedly one of the 13 major sources, by country, of greenhouse gas emissions (Marland, 1991). Not only are large quantities of natural gas flared in Nigeria but there is also extensive biomass burning, especially in the savanna zones. While reporting on the acidification potential in the Nigerian environment as part of the SCOPE Project, "Acidification in Tropical Countries," we had the opportunity to assess the contributions of emissions from biomass burning and energy use to acidification in Nigeria (Isichei and Akeredolu, 1988). We have also reported on the emissions of carbon (C), nitrogen (N), and sulphur (S) from biomass burning in Nigeria (Akeredolu and Isichei, 1991). For some aspects of these studies we had relied on earlier studies of N cycling in savanna ecosystems that had reported herbaceous biomass production and N content; the role of burning in N cycling in savannas was also assessed (Isichei and Sanford, 1980). For the present study of the fuel load characteristics for biomass burning assessment and emissions from burning, updated data from earlier studies (Isichei, 1983; Isichei, 1994; Isichei and Akeredolu, 1988; Isichei and Sanford, 1980) and more recent information from the literature were used to compile emissions of the greenhouse gases: carbon dioxide (CO_2), methane (CH_4), nitrogen dioxide (NO_2), carbon monoxide (CO), and other nitrogen compounds (NO_x), and nonmethane volatile organic hydrocarbon compounds.

2. Methodology of assessment of fuel load for biomass burning and resulting emissions

2.1. Savanna Burning

Most of the Nigerian savanna is burned annually. An estimated 20 700 × 10^3 ha (kha) of savanna is assumed to be subjected to annual fires. This is approximately half the area described as moist savanna. ERGO (1994) identified nine land-use types in its assessment of land-use change in Nigeria between 1976 and 1990. The categories were bare ground, grassland, scrub-shrubland, woodland, forest, mangrove, open water, cultivation, and settlement. The combinations of land-use types that are associated with savanna burning are farmland-woodland/shrub/grassland, wooded shrub grassland/woodland transition, and grassland wooded shrub.

FAO (1981) states that guinea savanna occupies 376 070 square kilometers (km^2), and the transition zone 41 560 km^2. These two zones, totalling 417 630 km^2, cover approximately 45% of Nigeria's land area. Our estimate of the annually burned area is approximately half the size of the guinea and transition zones. The ERGO land-use categories that burn annually are estimated to be 42 108 × kha. When farmland categories (22 246 × kha, according to ERGO) are subtracted from this area (farmlands are not burned annually), 19 863 × kha remain, similar to our estimate of 20 700 × kha.

A value of 3 Mg of dry matter per hectare (Mg dm ha^{-1}) was arrived at after considering the grass production figures shown in Table I. De Leeuw (1978) suggested a predictive relationship between total annual rainfall and maximum annual herbage standing crop. Thus: grassland primary production (kg ha^{-1}) = 2.36 × annual rainfall (mm) + 34.7. Most moist savanna locations would fall into the 3 Mg dm ha^{-1} yr^{-1} category if this equation were used. The fraction of total biomass actually burned is based on the results of Isichei and Sanford (1980) and Menaut *et al.* (1993).

Grasses, both annuals and perennials, are the dominant life form in the herb layer in moist savanna. Because they decline at the end of the season they are considered dead at the time of burning (Collins, 1977). When burning occurs before the grasses dry, only the leaves burn in some species. The proportion of leaves to other plant parts varies widely depending on the grass species. No data are available on the proportions of plant parts (Delmas, 1995 a,b).

The proportion of biomass oxidized based on the observations of Isichei (1994) was estimated to be 0.80. The observation was derived by burning grass in an aluminum tray and weighing the "ash" after burning and cooling. The mean weight of ash was 20% of the oven dry weight. A value of 0.47 as C fraction in biomass was adopted in this analysis (Susott *et al.*, 1995).

The N:C ratio was estimated to be 0.0007 (Isichei, 1994). There have been extensive analyses of N in Nigerian savanna grasses (Isichei, 1983). Mean dry season N concentration is approximately 0.35%, giving an N:C ratio of 0.007. The International Panel on Climate

TABLE I

Grass production figures for Nigeria

Grass standing crop at peak standing crop (Mg dm ha^{-1})	Reference
3.04	Isichei & Sanford (1980) (mean value for 9 sites)
0.8–6	Usman (1981)
1.5–7	Afolayan (1978)
2.01–4.5	De Leeuw (1976)
5.7–6.8	Hopkins (1966)
2.21–3.16	Muoghalu & Isichei (1991)

Note: Mg dm ha^{-1} = megagrams of dry matter per hectare.

Change (IPCC) default values for trace gas emission factors (for example, NO_x) and emission ratios do not differ significantly from the estimates of Delmas *et al.* (1995, this volume).

Burning is much less extensive in the Sudan region of Nigeria relative to the more humid savanna zones. However, there are no precise data on the areal extent of savanna burning in the Nigerian Sudan zone. We have adopted a value of 30% as the proportion of savanna burnt annually (Delmas *et al.*, 1991), and a herbaceous production figure of 2 Mg dm ha^{-1} yr^{-1}. Burning of agricultural residues is not common in Nigeria, as they are in great demand as livestock feed and construction material (Bayer and Otchere, 1985).

2.2 Land-use Change and Forestry

2.2.1. Forest clearing: CO_2 release from burning aboveground biomass on- and offsite
FAO (1986) reported an annual rate of deforestation of 300 000 ha for Nigeria, and this value was adopted in this analysis. The current extent of forests and plantations was obtained from NEST (1991) and Ojo (1994). For woodlands, deforestation was estimated to be 200 000 ha annually based on the observations of ERGO (1994). In their study, woodlands decreased from 44.63% to 41.3% between 1976 and 1990. When this decrease is converted to land area and divided by 15 years, a value of about 205 000 ha per year is obtained.

An average of 250 Mg dm ha^{-1} is used as forest dry matter, based on forest aboveground biomass of 229 Mg dm ha^{-1} for an undisturbed forest in southwestern Nigeria (Okali and Ola-Adams, 1987). Assuming a higher standing crop value for the more humid forests of southeastern Nigeria, a value of 250 Mg ha^{-1} was employed for undisturbed forests. For woodlands, a value of 50 Mg dm ha^{-1} was adopted based on prior estimates by Isichei (1994).

It is assumed, based on the work of Okali and Ola-Adams, that all fruit (0.4% of biomass), leaves (3%), and branches (33.3%) and 50% of stems (29.5% of biomass) are burned in savanna fires. The IPCC combustion default value of 0.9 was adopted for logged forest, because only branches and wood not considered useful are left behind, and these burn easily (UNEP *et al.,* 1995). Further, the IPCC value of 0.9 as fraction of biomass oxidized offsite was adopted, with most of the material used mainly for firewood. Approximately 24% of dry matter is assumed left to decay onsite. This includes tree stumps and large woody material not suitable for use as firewood and timber (which we assume constitute 10%). For the logged forest, a fraction of 0.1 was adopted, because 0.9 is assumed to be oxidized onsite.

Bebwa and Lejoly (1993) obtained a surface soil total C content of 16 Mg ha^{-1} in a 6-year fallow in Zaire. The value reported for the moist Lamto savanna in the Côte d'Ivoire was 17.8 Mg (Hall *et al.,* 1994). If a soil density of about 1 and an organic matter content of 3% (surface soil, 15-centimeter (cm) depth) for forest are assumed (Müller-Sämann and Kotschi, 1994), then a soil C content of 30 Mg ha^{-1} would be obtained.

2.2.2. Land-use change
ERGO (1994) reported that 2.3% of Nigeria's land area (920 000 km^2) is converted from grassland to cultivated land each year. This proportion equals 42 320 × kha, a value adopted as total conversion of grassland to cultivation over a 20-year period. A value of 15 Mg ha^{-1} was adopted as soil C content of grasslands (Müller-Sämann and Kotschi, 1994). The IPCC value of 2% was used in our analysis. Jones and Wild (1975) estimate an annual C mineralization rate of 4% in the savanna zones of West Africa, but this estimate is considered too high (see Müller-Sämann and Kotschi, 1994).

3. Results and discussion

3.1. Biomass Burning

Tables II and III show the stepwise estimates of emissions of C and non-CO_2 trace gases from biomass burning in Nigerian savanna. One crucial factor in making these estimates is the areal extent of biomass burning in Nigeria. Tekie (1989) estimates that up to 80% of natural savanna is burned in annual fires. The areal extent of burning savanna varies from year to year and from one location to the other, depending on such factors as land use and seasonality. Better estimates of the areal extent of savanna burning are now available (Scholes, this volume, and it is becoming clear that burning may have been overestimated in the past. Brustet *et al.* (1991) estimated CO_2 emissions from the burning of 180,000 km^2 of moist savanna in Côte d'Ivoire as 90 × 10^6 Mg for the months of December, January, and February. Using the same area estimate extent as in this study, Akeredolu and Isichei (1991) reported a CO_2 emission of 37 teragrams (Tg), 1.6 Tg of CH_4, and 0.0018 Tg of NO_X. Interpolation of Nigeria's latitude and longitude coordinates from the gridded emission data of Crutzen *et al.* yields a figure of 35 Tg CO_2 for savanna burning. These values are higher than those reported here, and it needs to be stated that estimates are improving as more refined estimates of savanna area burned, fraction of biomass burned, and emission ratios are available. Andreae *et al.* (1995) have concluded that N emissions from savanna burning are in the range of 14 – 33 kg ha^{-1}, corresponding to 60% of the aboveground N pool. Remote sensing offers the best method of estimating the areal extent of burning, both on a year-to-year basis and averaging over a number of years.

The season of burning, vegetation type and phenology are especially important because they affect the quantity of trace gases emitted from fires. Nutrients, especially the N content of savanna grass, drop exponentially toward the end of the growing season. As more areas become exotic tree plantations, more detailed information on the extent of each tree type and the frequency of fires under each species will be needed. Litter production under the various plantations is well known, but again, the areal extent of various plantation species such as *Gmelina, Tectona grandis, Eucalyptus,* and *Pinus* etc. are not exactly known.

TABLE II

Carbon released from biomass burning in Nigerian savanna

Biomass type	Area burned (kha)	Biomass density (Mg dm ha[-1])	Fraction burned	C released[1] (Gg)
Humid savannah herbaceous layer	20 700	3	0.3	23 350
Humid savannah litter	20 700	2	0.2	3 892
Plantation	43	7	0.9	127
Total C released				27 369

[1]Assumes biomass is 47% carbon

TABLE III

Non-CO_2 trace gases released from biomass burning in Nigerian savanna

Trace gas	Emissions (Gg)
CH_4 (emission ratio = 0.004, Molar ratio = 16/12)	145
CO (emission ratio = 0.06, Molar ratio = 28/12)	3 831
N_2O (emission ratio = N:C ratio = 0.007, Molar ratio = 44/28)	2
NO_x (emission ratio = 0.12, Molar ratio = 30/14)	49

3.2. Land-use Change and Forestry

3.2.1. CO_2 release from forest and woodland clearing

The relationships between forest clearing, forest and woodland burning on- and offsite, and C and CO_2 release are shown in Tables IV through VI. An estimated 125 561 Gg CO_2 is released from burning of cleared forest, delayed emissions from decay, and long-term emissions from soil in Nigerian forests and woodlands (Table VII).

3.2.2. CO_2 release from conversion of grassland to cultivated land

An additional 46 552 Gg CO_2 is estimated to be released from a 20-year conversion of grassland to cultivated land (Table VIII). This demonstrates the importance of fallow land, which is often not given as much attention as mature, preserved forests and woodlands in ecological studies.

Based on an N:C ratio of 0.005 (Isichei, 1994) in wood, 300 Gg CH_4, 4375 Gg CO, 2.4 Gg N_2O, and 24 Gg NO_X are estimated to be released from the burning of cleared forests and woodlands (Table IX).

TABLE IV

Biomass cleared annually from Nigerian forests and woodlands

Forest type	Area cleared (kha)	Biomass before (Mg dm ha^{-1})	Biomass after (Mg dm ha^{-1})	Annual loss (Gg dm)
Undisturbed	300	250	5	73 500
Logged	300	200	5	58 500
Woodland	200	50	3	9 400

TABLE V

Carbon released from onsite burning of biomass cleared from Nigerian forests and woodlands

Forest type	Biomass burned (Gg dm^{-1})	Fraction oxidized	C released[1] (Gg)
Undisturbed	18 375	0.66	5 700
Logged	14 625	0.9	6 186
Woodlands	2 350	0.8	884
Total C released			12 770

[1]Assumes biomass is 47% carbon

TABLE VI

Carbon released from off-site burning of biomass cleared from Nigerian forests and woodlands

Forest type	Biomass burned (Gg dm^{-1})	Fraction oxidized	C released[1] (Gg)
Undisturbed	7 350	0.9	3 109
Logged	5 850	0.9	2 474
Woodlands	940	0.9	398
Total C released			5 981

[1]Assumes biomass is 47% carbon

TABLE VII

CO_2 emissions from forest and woodland clearing in Nigeria

C from burning (Gg)	C from decay (Gg)	C from soil (Gg)	CO_2 from clearing[1] (Gg)
18 751	10 993	4 500	125 561

[1]Conversion ratio 44/12

TABLE VIII

CO_2 emissions from conversion of grassland to cultivated lands in Nigeria

A 20-year total conversion of grasslands to cultivation (kha)	B Soil carbon content of grasslands (Mg C ha^{-1})	C Annual rate of carbon release from soil	D Total annual soil carbon release from grassland conversion (Gg C)	E Total CO_2 released from historic conversion over 20 years (Gg CO_2)
			D=(AxBxC)	E=(Dx[44/12])
42 320	15	0.02	12 696	46 552

TABLE IX

Non-CO_2 trace gases from burning of biomass cleared from forests and woodlands in Nigeria

A Carbon Released (Gg C)	B Nitrogen-Carbon Ratio	C Total nitrogen Released (Gg N)		D Trace Gas Emission Ratios	E Trace Gas Emissions (Gg C)	F Conversion Factors	G Trace Gas Emissions From Burning Cleared Forests (Gg CH_4, CO)
					E=(A × D)		G=(E × F)
18 751	0.005	94	CH_4	0.012	225	16/12	300
			CO	0.1	1 875	28/12	4 375
					Gg N		Gg N_2O, NO_x
		C=(AxB)			E=(C × D)		G=(E × F)
			N_2O	0.007	0.658	44/28	2.4
			NO_x	0.121	11	30/14	24

4. Conclusion

The IPCC reporting format (UNEP *et al.*, 1995) requires a reliable national database of land-use emissions, which is absent for Nigeria. Most of the earlier estimates of emissions from burning and land-use changes are largely based on extrapolations (Crutzen *et al.*, 1990). Our emission values are lower, but are based on more recent and detailed information (Isichei, 1994).

References

Afolayan, T.A.: 1978, Grass biomass production in a Northern Guinea savanna ecosystem, *Oecologia Plantarum*, **13**, 375-386.

Akeredolu, F.A., and Isichei, A.O. 1991, Emissions of carbon, nitrogen and sulphur from biomass burning in Nigeria, in Levine, J.S. (ed.), *Global Biomass Burning*, MIT Press, 162-166.

Andreae, M.O., Atlas, E., Cachier, H., Cofer III, W.R., Harris, G.W., Helas, G., Koppmann, R., Lacaux, J.P., and Ward, D.E.: 1995, Trace gas and aerosol emissions from savanna fires, in Levine, J.S. (ed.), *Biomass Burning and Global Change*, MIT Press, in press.

Bayer, W., and Otchere, E.O.: 1985, Effect of livestock-crop integration on grazing time of cattle in a subhumid African savanna, in Tothill, J.C., and Mott, J.J (eds.), *Ecology and Management of the World's Savannas*, Australian Academy of Science, Canberra, 256-259.

Bebwa, B., and Lejoly, J.: 1993, Soil organic matter dynamics and mineral nutrients content in a traditional fallow system in Zaire, in Mulongoy, K., and Merckx, R. (eds.), *Soil Organic Matter Dynamics and Sustainability of Tropical Agriculture*., Wiley-Sayce, Chichester, 135-142.

Brown, S., Hall, C.A., Knabe, W., Raich, J., Trexler M.C., and Woomer, P.: 1993, Tropical forests: Their past, present, and potential future role in the terrestrial carbon budget, *Water, Air, and Soil Pollution* **70**, 71-94.

Brustet, J.M., Vickos, J.B., Fontan, J., Manissadjan, K., Podaire, A., and Lavenu, F.: 1991, Remote sensing of biomass burning in West Africa with NOAA-AVHRR, in Levine, J.S. (ed.), *Global Biomass Burning*, MIT Press, Cambridge, 47-52.

Collins, N.M.: 1977, Vegetation and litter production in Southern Guinea savanna, Nigeria, *Oecologia (Berlin)* **28**, 163-175.

Crutzen, P.J., Hao, W.M., Liu, M.H., Lobert, J.M., and Scharffe, D.: 1990, Emissions of CO_2 and other trace gases to the atmosphere from fires in the tropics: Our changing atmosphere, in Crutzen, P.J., Gerard, J.C., and Zander, R. (eds.), *Proceedings of the 28th Liege International Astrophysical Colloquium*, University of Liege, Cointre-Ougree, 449-472.

De Leeuw, P.N.: 1978, The primary and secondary productivity of the savanna in the middle belt of Nigeria. Paper presented at the Annual Conference of the Nigerian Society for Animal Production, Ahmadu Bello University, Zaria, Nigeria.

De Leeuw, P.N.: 1976, Fodder resources and livestock development in northeastern Nigeria, *Savanna* **5**, 61-74.

Delmas, R.A., Loudjiani, P., Poudaire, A., and Menaut, J.C.: 1991, Biomass burning in Africa, in Levine, J.S. (ed.), *Global Biomass Burning*, MIT Press, Cambridge, 126-132.

Delmas, R., Lacaux, J.P., and Brocard, D.: 1995a, Determination of biomass burning emission factors: methods and results, *Environmental Monitoring and Assessment*, this volume.

Delmas, R., Lacaux, J.P., Menaut, J-C., Abbadie, L., Le Roux, X., Helas, G., and Lobert, J.: 1995b, Nitrogen compound emission from biomass burning in tropical African savanna FOS/DECAFE 1991 Experiment (Lamto, Ivory Coast), *Journal of Atmospheric Chemistry*, in press.

ERGO (Environmental Research Group, Oxford): 1994, *Land Use Change in Nigeria, 1976-1990*, Federal Environmental Protection Agency, Abuja, Nigeria.

FAO (Food and Agricultural Organization of the United Nations): 1981, *Tropical Forest Resources Assessment Project: Forest Resources of Tropical Africa Country Briefs, Part II*. FAO, Rome.

FAO: 1986, *FAO Production YearBook*, FAO, Rome.

Frost, P.G.H.: 1985, Organic matter and nutrient dynamics in a broadleafed African savanna, in Tothill, J.C., and Mott, J.J. (eds.), *Ecology and Management of the World's Savannas*, Australian Academy of Science, Canberra, 200-206.

Hall, D.O., Scurlock, J.M.O., Ojima, D.S., and Parton, W.J.: 1994, Grasslands and the global carbon cycle: Modelling the effects of climate change, in *The Carbon Cycle: 1993 OIES Global Change Institute Proceedings,* Cambridge University Press, Cambridge, U.K.

Hopkins, B.: 1966, Vegetation of the Olokemeji Forest Reserve, Nigeria, IV: The litter and soil, with special reference to their seasonal changes, *Journal of Ecology* **54**, 687-703.

ILCA (International Livestock Centre for Africa): 1979, *Livestock Production in the Sub-Humid Zone of West Africa: A Regional Review,* ILCA Systems Study 2, ILCA, Addis Ababa.

Isichei, A.O.: 1983, Nitrogen concentration in the major grasses of the Derived and Guinea savanna zones of Nigeria in relation to season and site, *Tropical Agriculture* **60**, 48-52.

Isichei, A.O.: 1994, The stocks of nitrogen in vegetation and soil in West African moist savannas and the potential effects of climate change and land use on these stocks, *Global Ecology and Biogeography Letters,* in press.

Isichei, A.O., and Akeredolu, F.: 1988, Acidification potential in the Nigerian environment, in Rodhe, H., and Herrera, R. (eds.), *Acidification in Tropical Countries,* John Wiley & Sons, 297-316.

Isichei, A.O., and Sanford, W.W.: 1980, Nitrogen loss by burning from Nigerian grassland ecosystems, in Rosswall, T. (ed.), *Nitrogen Cycling in West African Ecosystems,* Royal Swedish Academy of Sciences, Stockholm, 325-331.

Jones, M.J., and Wild, A.: 1975, *Soils of the West African Savanna,* Commonwealth Bureau of Soils, Harpenden, U.K.

Keay, R.W.J.: 1959, *An Outline of Nigerian Vegetation* (3rd ed.), Government Printer, Lagos, Nigeria.

Marland, G.: 1991, Tropical forests and the greenhouse effect (commentary), *Climate Change* **19**, 227-232.

Menaut, J-C., Barbault, R., Lavelle, P., and Lepage, M.: 1985, African savannas: Biological Systems of Humification and Mineralization, in Tothill, J.C., and Mott, J.J. (eds.), *Ecology and Management of the World's Savannas,* Australian Academy of Science, Canberra, 14-33.

Menaut, J-C., Abbadie, L., and Vitousek, P.M.: 1993, Nutrient and organic matter dynamics in tropical ecosystems, in Crutzen, P.J., and Goldammer, J.G. (eds.), *Fire in the Environment: The Ecological, Atmospheric, and Climatic Importance of Vegetation Fires,* John Wiley & Sons, New York, 215-231.

Müller-Sämann, K.M., and Kotschi, J.: 1994, *Sustaining Growth: Soil Fertility Management in Tropical Smallholdings,* Margraf Verlag, Weikersheim, Federal Republic of Germany.

Muoghalu, J.I., and Isichei, A.O.: 1991, Man and biosphere research plots in Guinea savanna: Floristics and structure of the vegetation, *African Journal of Ecology* **29**, 229-240.

NEST (Nigerian Environmental Study /Action Team): 1991, *Nigeria's Threatened Environment: A National Profile,* NEST, Ibadan, Nigeria.

Ojo, L.O.: 1994, *Update of Wood Supply and Demand Data,* report submitted to FORMECU, Ibadan.

Okali, D.U.U., and Ola-Adams, B.A.: 1987, Tree population changes in treated rain forest at Omo Forest Reserve, Nigeria, *Journal of Tropical Ecology* **3**, 291-313.

Sanford, W.W., and Isichei, A.O.: 1986, Savanna, in Lawson, G.W. (ed.), *Plant Ecology in West Africa: Systems and Processes,* John Wiley and Sons, New York, 95-149.

Skarpe, C.: 1992, Dynamics of savanna ecosystems, *Journal of Vegetation Science* **3**, 293-300.

Susott, R.A., Baker, S.P., Olbu, G., Ward, D.E., and Kaufmann, J.B.: 1995, Carbon, hydrogen and nitrogen content of tropical ecosystem fuels, in Levine, J.S. (ed.), *Biomass Burning and Global Change,* MIT Press, Cambridge, in press.

Swaine, M.D.: 1992, Characteristics of dry forest in West Africa and the influence of fire, *Journal of Vegetation Science* **3**, 365-374.

Tekie, T.G.: 1989, *The Remote Sensing of Vegetation and Land Use in the Kainji Lake Basin, Nigeria,* Ph.D. thesis, University of London.

UNEP, OECD, IEA, IPCC (United Nations Environment Programme, Organization for Economic Co-operation and Development, International Energy Agency, Intergovernmental Panel on Climate Change): 1995, *IPCC Guidelines for National Greenhouse Gas Inventories,* IPCC, Bracknell, 3 Volumes.

Usman, S.S.: 1981, *Production of Herbaceous Material in the Guinea Savanna,* M.Sc. thesis, University of Ife, Nigeria.

Walker, B.H.: 1985, Structure and function of savannas: An overview, in Tothill, J.C., and Mott, J.J. (eds.), *Ecology and Management of the World's Savannas,* Australian Academy of Science, Canberra, 83-91.

White, F.: 1983, *The Vegetation of Africa: A Descriptive Memoir to Accompany the UNESCO/AETFAT/UNSO Vegetation Map of Africa,* UNESCO, Paris.

GREENHOUSE GAS EMISSION INVENTORY FOR SENEGAL, 1991

Y. SOKONA

Energy Program, Environmental Development
Action in the Third World, ENDA-B.P. 3370, Dakar, Senegal

Abstract. The first greenhouse gas (GHG) emission estimates for Senegal, for the year 1991, were produced according to the draft IPCC/OECD guidelines for national inventories of GHGs. Despite certain discrepancies, nonavailability of data, the quality of some of the data collected, and the methodology, the estimates provide a provisional basis for Senegal to fulfill its obligations under the UN Framework Convention on Climate Change. This inventory reveals that GHG emissions in Senegal, like those in many developing countries, can mainly be attributed to the use of biomass for energy, land-use change and forestry, and savanna burning. Taking into account the direct global warming potential of the main GHGs (CO_2, CH_4, and N_2O), Senegal's emissions are estimated at 17.6 Tg ECO_2. The major gases emitted are CO_2 (61% of GHG emissions), followed by CH_4 (35%) and N_2O (4%). Energy accounts for 45% of total emissions (12% from fossil energy and 33% from traditional biomass energy); land-use change and forests, 18%; agriculture, 24%; waste, 12%; and industry, 1%.

1. Introduction

To meet its obligations under the UN Framework Convention on Climate Change and to better understand the interaction between national concerns and global environmental problems, Senegal carried out a national inventory of its anthropogenic emissions of greenhouse gases (GHGs). The 1991 inventory included the following gases: carbon dioxide (CO_2), methane (CH_4), nitrous oxide (N_2O), nitrogen oxides (NO_X), and carbon monoxide (CO) (EPA, 1994). The base year was 1991 because the results of surveys and data collection on energy were available for that year. However, data on other sectors studied were available for 1989, 1990, and 1991. The relatively limited time (6 months) available for this work did not permit elaborate investigations of data and the emission coefficients required. However, the inventory did provide an opportunity to apply the IPCC/OECD guidelines, which were still being revised, in the context of Africa (IPCC/OECD, 1994).

2. Methodology

Greenhouse gas emissions were calculated using, to a very large extent, methods described in the three volumes of the IPCC/OECD *Draft Guidelines for Greenhouse Gas Emissions* (IPCC/OECD, 1994). The reporting instruction models were used to collect the primary data and present the final results. The default methods were used whenever a more relevant method could not be applied.

Environmental Monitoring and Assessment **38**: 291-299, 1995.

The five sectors (energy, industry, land-use change and forestry, agriculture, and waste) listed in the guidelines were taken into account but were adapted to the economic sectors of Senegal. Once the basic data relating to the different categories in each sector were collected and analyzed, simple specific surveys were undertaken to complement or clarify some of the data or available information. Remote sensing was the principal tool used to estimate the area exposed to savanna fires and to assess land-use changes due to forest clearing (Brown and Gaston, 1995).

As indicated in the IPCC/OECD guidelines, the energy emission inventory used the top-down methodology to obtain the carbon (C) balance (EPA, 1994). The idea is to first estimate the apparent consumption, or the balance between primary production, imports, exports, and stock variations and the various fossil fuels used in the country. Greenhouse gas emissions from the nonenergy industrial sector related to cement only were estimated based on clinker production. (Clinker is obtained from limestone quarries. The limestone is crushed and baked at 1500 °C.)

Three components were used to estimate GHG emissions from agriculture: livestock, savanna burning, and rice (*Oryza* L.) cultivation (de Jode, 1995). Results from a 1989–90 national aerial census completed by the *Centre de Suivi Ecologique* were used to estimate livestock numbers. The number of dairy cows was based on a survey by the Department for Livestock Farming, which estimated dairy cows to be 37.5% of the cattle population. Because this aerial census could not distinguish between sheep and goats, they were both combined under the heading small ruminants. Nor was it possible to classify the cattle according to age. Emissions from the anaerobic fermentation of animal waste were not taken into account.

Remote sensing was used to estimate the areas affected by savanna fires (Menant, 1993). Once the different types of savanna were identified, the biomass density of each, as well as the combustible biomass, were determined directly from satellite images (an index map of the integrated vegetation at the end of the growth season), which were validated by onsite measurements. Contrary to the approach recommended by IPCC/OECD, which involves multiplying the quantity of biomass per hectare by the area burned, biomass data were obtained by superimposing the vegetation production map onto the fire map. This allows the area burned to easily be determined. The part of the woody biomass not burned were not taken into account.

The low quantities of agricultural residues burned on site were estimated, although no distinction was made between millet (*Setaria italica*) and sorghum (*Sorghum vulgare* L.). The residue/crop product ratio was estimated based on discussions with experts in specific sectors. The emissions from waste were calculated from the results of a specially conducted survey and according to the approach suggested in the IPCC/OECD guidelines.

Two components were considered for the land-use change and forestry sectors: forest clearing and managed forests. To calculate the emission share of each component, the country was divided into three broad ecological categories: woody savanna, savanna, and open forest; scrublands; and grassy scrublands. It was assumed that forests were cleared for agriculture only in the woody savanna, savanna, and open forest category. This assumption was mainly based on the 1991 Forestry Development Master Plan and the 1993

Forestry Action Plan. Conversely, managed forestry was applied to each of the three vegetation categories. A survey on woodfuel consumption was used to determine the different parameters needed to calculate GHG emissions.

3. Results and discussion

3.1. Overview

The energy sector accounts for 45% of total GHG emissions, of which 12% comes from fossil fuels and 33% from traditional biomass fuels. Agriculture accounts for 24%, while land-use changes and forests, waste, and industry account for 18%, 12%, and 1%, respectively (Figure 1). A more detailed picture of the sectors, activity data, and GHG emissions included in emissions inventory is provided in Figure 2.

3.2. Carbon Dioxide (CO_2)

Eighty-nine percent of CO_2 emissions comes from biomass burning and land-use changes, approximately 10% comes from fossil fuels, and almost 1% comes from cement production

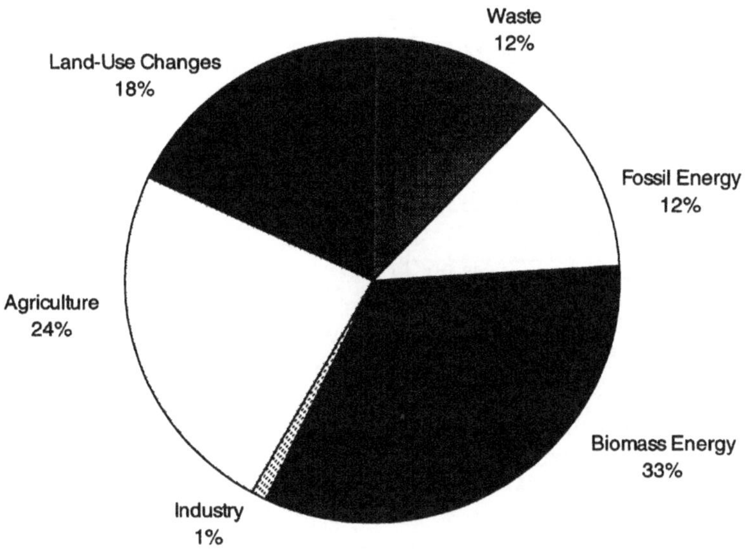

Figure 1. Proportion of GHG emissions from major economic sectors.

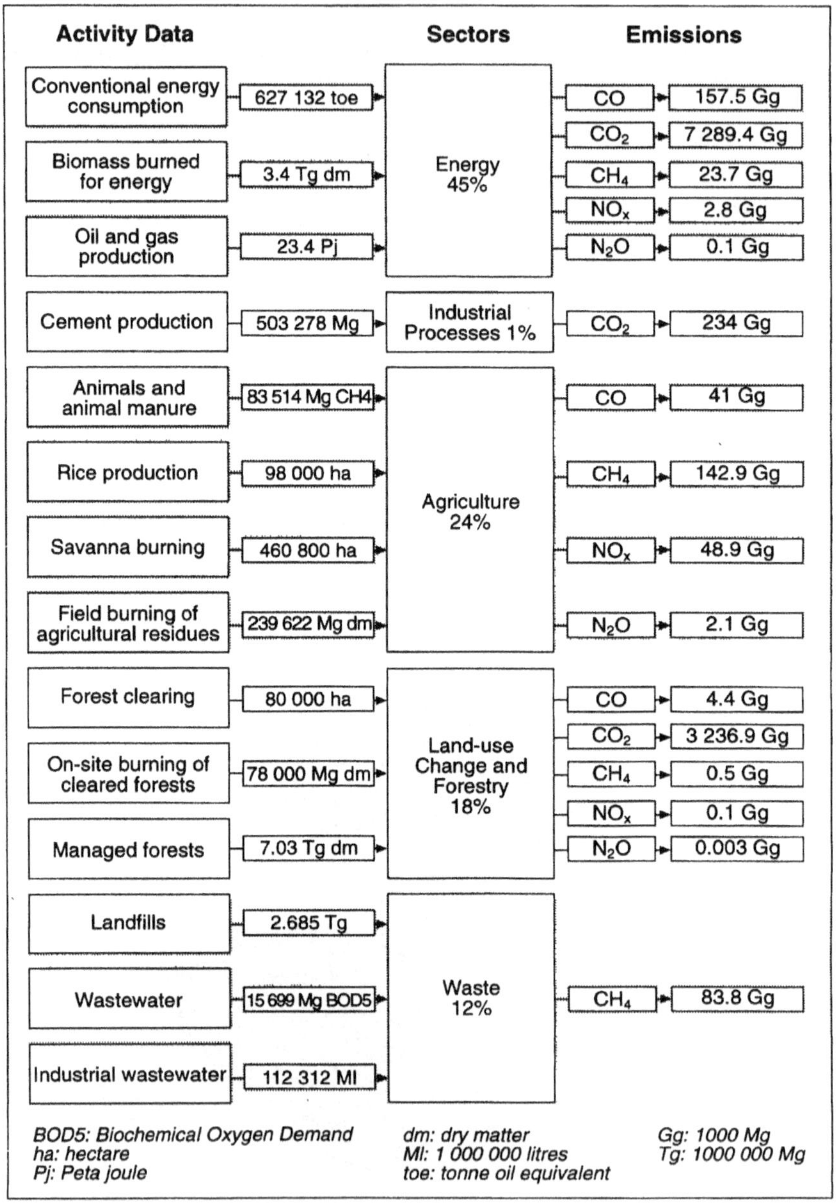

Figure 2. Activity data, economic sectors, and GHG emissions in Senegal, 1991.

(Figure 3). About 52% of CO_2 emissions is sequestered by CO_2 sinks stemming from forest activity, reforestation, and land-use practices. Table I shows the amount from each CO_2 source and sink.

CO_2 emissions from land-use change and forestry were calculated from estimates of various parameters given in the IPCC/OECD guidelines, because no reliable inventory of agriculture and forestry in Senegal had been completed for the reference period. The various data used were estimated from recent studies undertaken for the forestry sector. However, the available information does not distinguish between forest clearing for additional agricultural land and other types of ecosystem degradation (Singh, 1993). The greater portion of CO_2 emissions attributed to land-use changes confirms the shrinking of forest areas and the degradation of the land needed for the activities of the Senegalese forest industry.

Because there were no specific primary data on energy consumption in Senegal, only aggregate CO_2 emissions linked to energy were estimated. To calculate these emissions, energy data from various sources were collected. Emission coefficients for the various forms of energy within the Senegalese energy system were used in the calculations. The CO_2 emissions from industry related to the cement industry only, because it was the only industry for which data were available.

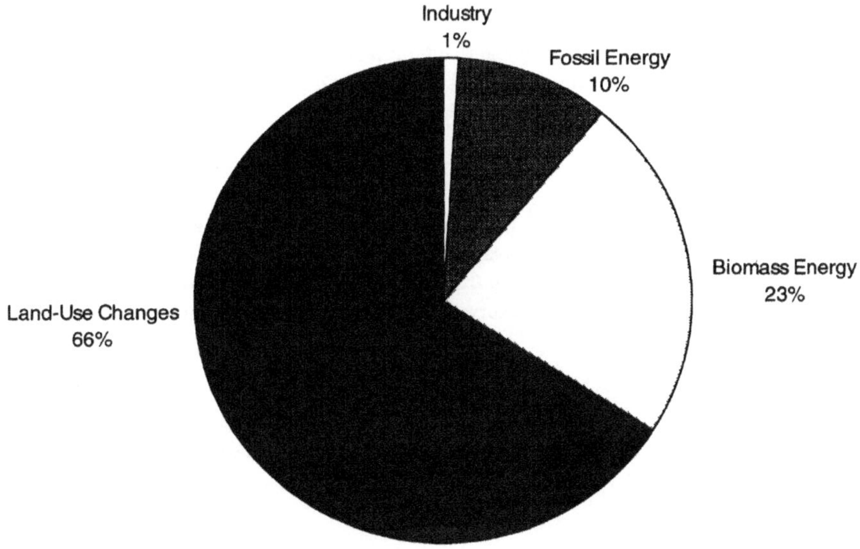

Figure 3. Distribution of CO_2 emissions sources.

TABLE I

Sources and sinks of CO_2 in energy and land-use sectors of Senegal, 1991 (Gg)

Category	Amount
Source	
Energy sector	7,289.4
Fossil fuels	2,192
Traditional biomass burning	5,097.4
Land-use change & forestry	14,832.8
Industry (cement)	234
Total emissions	22,356.2
Sink	
Forestry/reforestation	-11,595.8
Total CO_2	10,760.4

Efforts to reforest and improve agricultural practices help sequester CO_2 in biomass. However, these sinks sequester less CO_2 than the CO_2 emissions from the expansion of agricultural land or woodfuel consumption.

3.3. Methane (CH_4)

Agriculture accounts for more than half (57%) of the CH_4 emissions, mainly from commercial livestock and rice cultivation (Table II). About 33% comes from waste, a little over 9% from energy, and less than 0.5% from land-use change and forestry. The basic data were estimated from specific surveys and from default values indicated in the IPCC/OECD manual.

3.4. Nitrous Oxide (N_2O)

N_2O emissions are relatively minor in Senegal. Table III shows the estimated amount of nitrous oxide emitted from energy, agriculture and land-use changes. Agriculture is the greatest NO_2 source.

TABLE II

Methane emissions in Senegal from various sources, 1991 (Gg)

Source	Amount emitted
Energy	23.8
Combustion	23.7
Fugitive emissions	0.1
Agriculture	142.9
Livestock farming	83.5
Rice	58.7
Waste burning	0.7
Land-use changes	0.5
Waste	83.8
Rubbish collection	68.9
Industrial wastewater	14.8
Total	251

3.5. Carbon Monoxide (CO) and Nitrogen Oxides (NO_x)

Table IV shows the estimated amount of CO and NO_x emissions from energy, agriculture and land-use changes. Energy is largest source of CO and agriculture is largest source of NO_x.

TABLE III

Nitrous oxide emissions in Senegal from three sources, 1991 (Gg)

Source	Amount emitted
Energy	0.1
Agriculture	2.1
Land-use changes	0.003
Total	2.203

TABLE IV

Precursor (CO, NO_x) emissions in Senegal from three sources, 1991 (Gg)

Source	CO	NO_x
Energy	157.5	2.8
Agriculture	41	48.9
Land-use change & forestry	4.4	0.1
Total	202.9	51.8

3.6. Global Warming Potential (GWP)

Different greenhouse gases vary in the effectiveness with which they trap heat energy in the atmosphere. In other words, similar quantities of different greenhouse gases have different global warming potentials (GWPs). These differences are usually expressed relative to the effectiveness of CO_2. For example, a molecule of CH_4 is 24.5 times more effective in warming the atmosphere than a molecule of CO_2 and a molecule of N_2O is 320 times more effective than a molecule of CO_2 (Houghton *et al.*, 1992). When account is taken of the GWPs of the main greenhouse gases (CO_2, CH_4, and N_2O) Senegal's greenhouse gases amounted to 17.6 teragrams (Tg) on a CO_2 equivalent basis.

Accounting for the global warming potential (GWP), emissions of primary greenhouse gases (CO_2, CH_4, and N_2O) come to a total of 17.6 Tg ECO_2 (equivalent CO_2). Carbon dioxide represents 61% of the total, CH_4 about 35%, and N_2O 4%.

The energy sector accounts for 45% of total GHG emissions, of which 12% comes from fossil fuel and 33% from traditional biomass energy. Land-use change and forests account for 18% of total GHG emissions; agriculture, waste, and industry account for 24%, 12%, and 1%, respectively.

4. Discussion and conclusions

Senegal's GHG emissions are approximately 18 Tg annually. Thus, Senegal is not a major GHG producer in Africa or at a global scale (Houghton *et al.*, 1992). The leading GHGs in Senegal are CO_2, CH_4 and N_2O accounting for 61, 35 and 4%, respectively, of annual emissions. Energy consumption, land-use change, and agricultural practices are leading sources of GHG emissions. These emissions trends and sources are similar to other sub-Sahelian countries (e.g., Ivory Coast).

Although the methodology applied has some deficiencies, the final GHG inventory provides useful emissions data to identify mitigation options (Rep. de Senegal, 1994), and meet commitments under UN Framework Convention on Climate Change (Callander 1995). Improved data on forest inventory (Singh, 1993), agricultural systems (de Jode, 1995), and the energy sector (EPA, 1994) are needed to refine the GHG inventory. Some assumptions, especially default values, should be adjusted for the socioeconomic conditions of developing countries.

References

Brown, S. and Gaston, G.: 1995, Use of forest inventories and geographic information systems to estimate biomass density of tropical forests: application to tropical Africa. Environmental Monitoring and Assessment, in press

Callander, B.: 1995, Scientific Aspects of the Framework Convention on Climate Change and National Greenhouse Gas Inventories, Environmental Monitoring and Assessment, in press.

de Jode, A.: 1995, Assessing national livestock populations for the production of methane emissions inventories. Environmental Monitoring and Assessment, in press.

Environmental Protection Agency—Office of Policy, Planning, and Evaluation (EPA-OPPE): 1994, *Inventory of U.S. Greenhouse Gas Emissions and Sinks: 1990–1993.*

Houghton, J.T., Callander, B.T., and Varney, S.K. (eds.): 1992, The Supplemental Report to the IPCC Scientific Assessment, IPCC, Cambridge, U.K.: Cambridge University Press.

IPCC/OECD (Intergovernmental Panel on Climate Change/Organisation for Economic (Co-operation and Development) Joint Programme: 1994, *IPPC Draft Guidelines for National Greenhouse Gas Inventories,* IPCC/OECD Joint Programme, Paris, 3 volumes.

Menant, J.C.: 1993, Effect des feux de savane sur le stockage et l'emission du carbone et des elements trace, *Revue Secheresse.*

Republic de Senegal—Ministere de l'Environment et de la Protection de la Nature: 1994, Inventaire des emissions de gaz a effet de serre au Senegal.

Singh, K.D. 1993, Forest Resources Assessment 1990, FAO Forestry Paper 112, FAO, Rome, Italy.

EMISSIONS OF GREENHOUSE GASES FROM AGRICULTURE, LAND-USE CHANGE, AND FORESTRY IN THE GAMBIA

B.P. JALLOW

Principal Meteorologist, Department of Water Resources,
7, Marina Parade, Banjul, The Gambia

Abstract. The Gambia has successfully completed a national greenhouse gas emissions inventory based on the results of a study funded by the United Nations Environment Programme (UNEP)/Global Environment Facility (GEF) Country Case Study Program. The concepts of multisectoral, multidisciplinary, and interdisciplinary collaboration were most useful in the preparation of this inventory. New data were gathered during the study period, some through regional collaboration with institutions such as Environment and Development in the Third World (ENDA–TM) Energy Program and the Ecological Monitoring Center in Dakar, Senegal, and some through national surveys and the use of remote sensing techniques, as in the Bushfires Survey. Most of the data collected are used in this paper. The Intergovernmental Panel on Climate Change/Organisation for Economic Co-operation and Development/International Energy Agency (IPCC/OECD/IEA) methodology is used to calculate greenhouse gas emissions. Many of the default data in the IPCC/OECD/IEA methodology have also been used. Overall results indicate that in the biomass sectors (agriculture, forestry, and land-use change) carbon dioxide (CO_2) is emitted most, with a total of 1.7 Tg. This is followed by methane (CH_4), 22.3 Gg; carbon monoxide (CO), 18.7 Gg; nitrogen oxides (NO_x), 0.3 Gg; and nitrous oxide (N_2O), 0.014 Gg. The Global Warming Potential (GWP) was used as an index to describe the relative effects of the various gases reported here. Based on the emissions in The Gambia in 1993, it was found that CO_2 will contribute 75%, CH_4 about 24.5%, and N_2O 0.2% of the warming expected in the 100-year period beginning in 1993. The results in this analysis are limited by the shortcomings of the IPCC/OECD/IEA methodology and scarce national data. Because the methodology was developed outside of the developing world, most of its emissions factors and coefficients were developed and tested in environments that are very different from The Gambia. This is likely to introduce some uncertainties into the results of the calculations. Factors and coefficients that are country- or region-specific are likely to provide more accurate results and should be developed. The surveys were conducted either during the wet season or just at the end of the wet season. This seasonal factor should contribute to variations in the results, particularly in the livestock numbers and composition survey. Use of one-time survey data is also likely to introduce uncertainty into the results.

1. Introduction

The Gambia has a "sudano-sahelian" type of climate, with a short rainy season from June to October and a long dry season from November to May. Mean annual temperature is about 28°C, and mean annual rainfall is about 850 mm. Over the past 40 years, there has been a slight warming trend and a 25 to 30% decrease in rainfall.

The population of The Gambia exceeds one million people, according to the April 1993 census. From 1983 to 1993 the population increased by about 49%, with an annual growth rate of 4.1 percent. The population density is among the highest in Africa, at 96 persons per square kilometer in 1993. Biomass fuels account for about 85% of the energy usage in the country (EU/MNRE, 1992). Agricultural activities contribute directly to the emission of greenhouse gases through a variety of processes; including methane (CH_4) emissions from enteric fermentation in domestic animals, animal wastes, and rice production.

Environmental Monitoring and Assessment **38**: 301-312, 1995.

Land-use changes that alter the amount of biomass on the land produce a net exchange of greenhouse gases between the atmosphere and land surface. The primary land-use changes that result in greenhouse gas emissions and uptake are conversion of forests to nonforests (for example, conversion of forests to pasture or cropland) and conversion of nonforests to forests (for example, establishment of plantations).

When forests are cleared, most of the C in the cleared biomass is released to the atmosphere as CO_2. Burning releases other gases in addition to the CO_2 as byproducts of incomplete combustion. These include CH_4, carbon monoxide (CO), N_2O, and nitrogen oxides (NO_x). Emissions of CO_2 from land clearing may or may not imply a net release of CO_2 to the atmosphere, but emissions of the other trace gases are net transfers from the biosphere to the atmosphere.

Land-use changes also produce greenhouse gas emissions through disturbance of forest soils. When forests are converted to croplands, an average of about 25 to 50% of the soil C is released as CO_2, primarily through oxidation of organic matter (UNEP *et al.,* 1995). Loss of forest may also result in increased net CH_4 emissions to the atmosphere because forest soils are a natural CH_4 sink. Forest conversion also releases N_2O.

The objective of this study is to characterize emission of greenhouse gases from agriculture, land-use change, and forestry sectors. Greenhouse gas emissions were estimated according to the *IPCC Guidelines for National Greenhouse Gas Inventories* (UNEP *et al.,* 1995). Emission estimates are summarized in Table I. Specific methods and assumptions are presented below.

2. Agriculture

2.1. Overview of the Agriculture Sector

Agriculture in The Gambia is entirely rainfed, and over 75% of The Gambia's labor force is engaged in subsistence farming. Agricultural technology is still rudimentary and extremely labor intensive. Fertilizer use in The Gambia averaged 25 kg ha in 1989, which is higher than in other West African countries but still 75% below recommended usage (Jabara, 1990).

2.2. Methane Emissions from Animals and Animal Manure

Both ruminant and some nonruminant animals (for example, pigs and horses) produce CH_4, although ruminants are the largest source. The amount of CH_4 released depends on the animal category, age, and weight, the quality and quantity of feed, and the energy expenditure of the animal.

2.2.1. Methodology and data used to estimate of emissions from animals and animal manure

Methane emissions from animals and animal manure were estimated as follows:

CH_4 enteric fermentation (EF)	=	number of animals per category (head counts) × emissions factor for enteric fermentation (per head/year).
CH_4 emissions from animal manure (AM)	=	number of animals per category (head counts) × emissions factor for manure management (per head/year) (Mg yr^{-1}).
Total methane emissions from livestock	=	EF + AM.

Data used in estimating emissions from animals and animal manure were obtained from the livestock numbers and composition survey conducted by the Department of Livestock Services (DLS) and the 1993 Statistical Year Book of The Gambia Agriculture, published by the Department of Planning (DoP) of the Ministry of Agriculture and Natural Resources.

2.3. Methane Emissions from Rice Production

In flooded rice fields, anaerobic decomposition of organic material by methanogenic bacteria produces CH_4, which escapes to the atmosphere primarily by diffusive transport through the rice plants (IPCC/OECD, 1994). Minor amounts of CH_4 escape via bubbles rising through the water column and by diffusion across the water/air interface. Experiments have shown that CH_4 fluxes vary with soil type, with temperature, with the type, timing, application mode, and amount of fertilizer applied, with water depth, with time of day, and with season.

2.3.1. Methodology and data used to estimate emissions from rice production

Methane emissions from rice production were estimated as follows:

CH_4 emissions from rice production (Gg)	=	harvested area (Mha) per water management regime × season length (days) × emission factor (kg/ha-day).

Data on area harvested were obtained from the 1993 Statistical Year Book of The Gambia Agriculture. Harvested area is defined as the physical area under cultivation times the number of harvests.

2.4. Emissions of Non-CO_2 Trace Gases from Savanna Burning

Savannas are tropical and subtropical vegetation formations characterized by a predominantly continuous grass cover, occasionally interrupted by trees and shrubs. Most vegetation growth occurs during the wet season. During the dry season, the grasses wither and die, and fires are frequent. Most of the fires are ignited humans.

Savanna burning produces instantaneous emissions of CO_2, but because most of the grasses regenerate during the following wet season, it is reasonable to assume that the net CO_2 released to the atmosphere is in equilibrium. In addition to CO_2, savanna burning releases other gases that are emitted as a result of incomplete burning and other varying factors such as ambient temperature. These gases include CH_4, CO, N_2O, and NO_x. Unlike CO_2 emissions from the burning of savanna, emissions of these other gases are net transfers from the biosphere to the atmosphere.

2.4.1 Methodology and data used to estimate emissions of non-CO_2 trace gases from savanna burning

The methodology used in estimating emissions of non-CO_2 trace gases from the burning of savanna involves determining the total C released from burning dead and living biomass and the use of N/C ratios and emissions factors to partition the total C released into the various trace gas emissions. This is expressed mathematically as:

Total C released	=	total biomass oxidized × (C fraction of living biomass + C fraction of dead biomass),
where total biomass oxidized	=	Area burned by category × biomass density of savanna × fraction actually burned × (fraction of living biomass burned + fraction of dead biomass burned).

The N content of the total C released is determined as:

N content	=	total C released × N/C ratio.

From the total C released, CH4 and CO are determined as:

CH_4 emissions	=	total C released × emissions ratio (0.004) conversion factor (16/12).
CO emissions	=	total C released × emissions ratio (0.060) conversion factor (28/12).

From the total N content N_2O and NO_x emissions are determined as:

N_2O emissions	=	total N content × emissions ratio (0.007) × conversion factor (44/28).
NO_x emissions	=	total N released × emissions ratio (0.120) × conversion factor (30/14).

The data used in the estimation of emissions of non-CO_2 trace gases from savanna burning were obtained from the Savanna Burning Survey (Annex IV) conducted by the Ecological Monitoring Center in Dakar. Heat-sensitive channels on the Advanced Very High Resolution Radiometer (AVHRR) apply to forest and agroecosystems because of their ability to detect fires. Fires are detected by the infrared radiation they emit. Fire temperature is usually around 500 to 600°C, with a peak emittance of 5 micrometers (Prince *et al.*, 1990). Because the temperature of the fire relative to the surrounding area is high, it can be detected by low spatial resolution systems such as AVHRR.

The process of determining the area and quantity of biomass burned involves the acquisition, correction, and classification (using Boxclass Module of CHIPS) of the satellite images. A false color composite (FCC) of the images is developed by loading the first visible channel (AVHRR Channel 1) to the blue gun of the monitor, the second visible channel (AVHRR Channel 2) to the green gun, and the near-infrared channel (AVHRR Channel 3) to the red gun. The burned areas are then discernible on the image as distinct dark spots that can be picked up in all subsequent images by the same method of classification and false color compositing. To determine the total area burned, the total number of pixels was counted and multiplied by one square kilometer, which is the area of one pixel. To determine the total biomass combusted, the resulting burned-surface map was overlaid by a biomass map of The Gambia, using the EXTRACT Module of the IDRISI geographic information system (GIS) software. The results showed the total area burned to be 35 500 ha for the 1991–1992 dry season, representing 60.4 Gg of dry matter destroyed by fire.

2.5. Emissions of Non-CO_2 Trace Gases from Field Burning of Agricultural Residues

Farming systems produce large quantities of agricultural wastes and burning these wastes is a common practice in the developing world. Residues are burned primarily to clear remaining straw and stubble after harvest and to prepare the field for the next cropping cycle. It is estimated that up to 40% of the residues produced in developing countries may be burned in fields, on the order of 425 Tg of dry matter agricultural wastes, or approximately 200 Tg C (IPCC/OECD, 1994). Like the burning of savannas, the burning of crop residues is not thought to be a net source of CO_2 emissions, because the C released to the atmosphere during burning is reabsorbed during the next growing season. However, crop residue burning is a significant source of CH_4, CO, NO_x, and N_2O.

2.5.1. Methodology and data used to estimate emissions from the burning of agricultural residues

The methodology used to estimate the emissions of trace gases from the burning of agricultural residues was the same used in estimating emissions from savanna burning. The data used in this study were taken from a survey conducted by the Department of Planning (DoP) of the Ministry of Agriculture and Natural Resources. Based on a sample survey of 222 villages, the objective of the survey was to estimate the quantity of crop residues burned in the field. Three families per village were selected for the survey. After sample selection, the head of the family was interviewed about agricultural residue use. Results of the survey indicate that 4% of the millet, 3% of the sorghum *(Sorghum* L.), 5% of the rice (*Oryza* L.), 6% of the maize (*Zea mays* L.), 8% of the cotton (*Gossypium*), and 1% of the groundnuts residues generated were burned in the field. However, because the sample frame for this survey was too small to be representative, additional data and information from national inventories within Africa and default data from (UNEP *et al.*, 1995).

3. Land-use change and forestry

3.1. Overview of the Land-use Change and Forestry Sector

It is estimated that forests cover about 43% of the land area of The Gambia, classified into four broad categories: closed forest (26 800 ha), open forest (62 600 ha), tree and shrub savanna (347 000 ha), and mangroves (66 900 ha) (DoF, 1991). The tree and shrub savanna woodland, which constitutes about 70% of all forest types in the country, has less than 10% tree cover or trees that are less than 11 meters high (Forster, 1983).

Forestry's share of the gross domestic product is estimated at less than 1%. However, this does not take into consideration the significant informal trade in timber and non-timber forest products (fuelwood, fencing posts, wood carvings, honey, palm oil and kernel, and wild fruits) that occurs locally in the rural areas and across the border. Forests provide more than 85% of the domestic energy needs of the country, in the form of woodfuel, and about 17% of the domestic timber needs (EU/MNRE, 1992).

3.2. CO_2 Emissions from the Burning of Aboveground Biomass On- and Off-site as a Result of Forest Clearing

Forests are usually cleared for permanent cropland or pasture by cutting the undergrowth and felling trees and then burning the trees on-site or as fuelwood. By this process, some of the biomass is burned while some remains on the ground, where it decays slowly (usually over a period of 10 years in the tropics). Of the burned material, a small fraction (5 to 10%) is converted to charcoal, which resists decay for 100 years or more, and the remainder is released instantaneously into the atmosphere as CO_2. Carbon is also lost from the forest soils after clearing, particularly when the land is cultivated. These emissions occur over periods of 25 years or more.

3.2.1. Methodology and data used in the estimates of emissions of CO_2 from forest clearing

The methodology used in estimating emissions of CO_2 from forest clearing involves determining total C released from burning of aboveground biomass both on- and off-site, from decay of biomass, and from soil C. The total C released is then multiplied by the conversion factors to obtain the total CO_2 released.

Total annual CO_2 from clearing	=	(total C released on-site + total C released off-site + total C from decay of biomass + total C forest from disturbance of soil C) × conversion factor of (44/12),

where

total C released on-site	=	area cleared annually × (biomass before clearing − biomass after clearing) × fraction of biomass exposed to burning on-site × combustion efficiency × C fraction of aboveground biomass (burned on-site),
total C released off-site	=	area cleared annually × (biomass before clearing − biomass after clearing) × fraction of biomass exposed to burning off-site × combustion efficiency × C fraction of aboveground biomass (burned off-site),
total C released due to decay	=	annual area cleared (10 year average) × (biomass before release from clearing − biomass after clearing) × fraction left to decay × C decay fraction in aboveground biomass, and

total soil C released = average annual forest cleared (10 year average) × soil C content of cleared land × fraction of C released.

The data used in this element of the study were taken from the results of the 1982 Forest Resources Inventory of The Gambia (Forster, 1983). Default activity data and emission ratios were also used (UNEP *et al.*, 1995). The Forest Resources Inventory was based on sampling, with pre-stratification conducted according to stand density; that is, the forests were classified according to the stock density (Forster, 1983). The pre-stratification was based on aerial photography data taken in October-November 1980. The number of samples taken within a land-use class depended on the expected forest volume and on the total area of land-use class. Thus, the number of samples was related to the total volume of the land use.

3.3. Non-CO_2 Trace Gas Emissions from On-site Burning of Cleared Forest

On-site burning of cleared forests resembles the other biomass-burning activities, such as burning of traditional biomass fields (energy sector), savanna burning, and field burning of crop residues (agriculture sector).

3.3.1. Methodology and data used to estimate of non-CO_2 trace gas emissions from on-site burning of cleared forests

To determine non-CO_2 trace gas emissions from on-site burning of cleared forests, the quantity of C released on-site is used as shown in section 3.2. The N content of the released C is determined so the N_2O and NO_x emissions can be estimated.

The N content of the total C released is determined as:

N content = total C released × N/C ratio.

From the total C released, CH_4 and CO are determined as:

CH_4 emissions = total C released × emissions ratio (0.004) × conversion factor (16/12), and

CO emissions = total C released × emissions ratio (0.060) × conversion factor (28/12).

From the total N content, N_2O and NO_x emissions are determined as:

N_2O emissions = total N content × emissions ratio (0.007) conversion factor (44/28).

NO_x emissions = total N released × emissions ratio (0.120) × conversion factor (30/14).

The data used were obtained from the 1982 Forest Resources Inventory of The Gambia.

3.4. CO_2 Emissions from Managed Forests

Managed forests include all categories that experience periodic or ongoing human intervention that affects C stocks. This also includes some tree planting activities, plantation management, and other silvicultural treatments.

3.4.1. Methodology and data used to estimate CO_2 emissions from managed forests
The annual CO_2 released, or CO_2 uptake, from managed forests was estimated as follows:
Total CO_2 = (Annual C released – total C increment) × 44/12,
where

annual C	=	(total biomass consumption – wood removed from forest clearing) × C fraction,
total C increment for managed forests	=	area of managed forests × annual growth rate × C content of dry matter, and
total C increment for afforestation programs	=	number of trees planted × annual growth rate × C content of dry matter

4. Results and conclusions

Table I and Figure 1 reveal CO_2 emissions are greatest in agriculture, land-use, and forestry subsectors, with a total of 1647.594 Gg, followed by CH_4 (22.255 Gg), CO (18.687 Gg), NO_x (0.335 Gg), and N_2O (0.014 Gg). Land-use change and forestry are the greatest sources of greenhouse gas emissions in the biomass sectors in The Gambia (Table I and Figure 2). Considering the size of The Gambia, net emissions of more than 1.6 Tg CO_2, about 14.5 Gg CO, and 1.7 Gg CH_4 per year from land-use change and forestry should be viewed as significant. Emissions of 1647.594 Gg CO_2 are the difference between emissions of 1661.196 Gg CO_2 from forest clearing and an uptake of 13.602 Gg of CO_2 from managed forests (plantations). These results suggest a need to identify and implement changes in land-use policy, law, and tenure that would discourage deforestation. These changes would significantly influence the distribution of resources among forestry, rangeland, and agriculture to encourage sustainable development. It would be useful to investigate mitigation objectives for non-greenhouse gas emissions including preservation and enhancement of the biodiversity of flora and fauna, soil conservation, and watershed management. Human activity and dependence on land resources should also be considered in future analyses.

4.1. Global Warming Potential

Greenhouse gases vary in their effectiveness to trap heat in the atmosphere. For example, a molecule of N_2O is 320 times more effective in warming the lower atmosphere than a molecule of CO_2. Similarly, a molecule of CH_4 is 24.5 times more effective in warming

TABLE I

Overview of greenhouse gas emissions from the biomass sectors in The Gambia in 1993

	Gigagrams (Gg)				
	Carbon Monoxide CO	Carbon Dioxide CO_2	Methane CH_4	Nitrogen Oxides NO_x	Nitrous Oxides N_2O
MODULE 4: AGRICULTURE					
Methane Emissions from Animals and Animal Manure			12.771		
Methane Emissions from Rice			7.662		
Savanna Burning, Release of Non-CO_2 Trace Gases	3.365		0.128	0.037	0.002
Field Burning of Agricultural Residues Release of Non-CO_2 Trace Gases	0.852		0.041	0.030	0.001
Subtotal Module 4	4.217		20.602	0.067	0.003
MODULE 5: LAND-USE CHANGE and FORESTRY					
Forest Clearing		1661.196			
On-site Burning of Cleared Forest	14.470		1.654	0.268	0.011
Managed Forests		−13.602			
Subtotal Module 5	14.470	1647.594	1.654	0.268	0.011
Total	18.687	1647.594	22.255	0.335	0.014
Global Warming Potential (GWP), 100 years integration		1.0	24.5		320.0
1000 TCO_2E		1647.594	545.255		4.551
Percent		75.0	24.8		0.2

the lower atmosphere than a molecule of CO_2 (Houghton *et al.,* 1992). Adjusting for this difference, Figure 3 compares emissions of CO_2, CH_4, and N_2O on a CO_2-equivalent basis. Based on the emissions of these gases in the biomass sectors of The Gambia in 1993, CO_2 will contribute 75%, CH_4 about 24.8%, and N_2O 0.2% to the warming expected in the 100-year period beginning in 1993 (Figure 4).

This method of comparing emissions of trace gases is referred to as global warming potential (GWP). It is the time-integrated warming effect of the instantaneous release of a unit mass of a given greenhouse gas in today's atmosphere, relative to the release of CO_2. The index thus describes the relative effectiveness of various greenhouse gases in contributing to potential global warming. Considering the GWP of each greenhouse gas

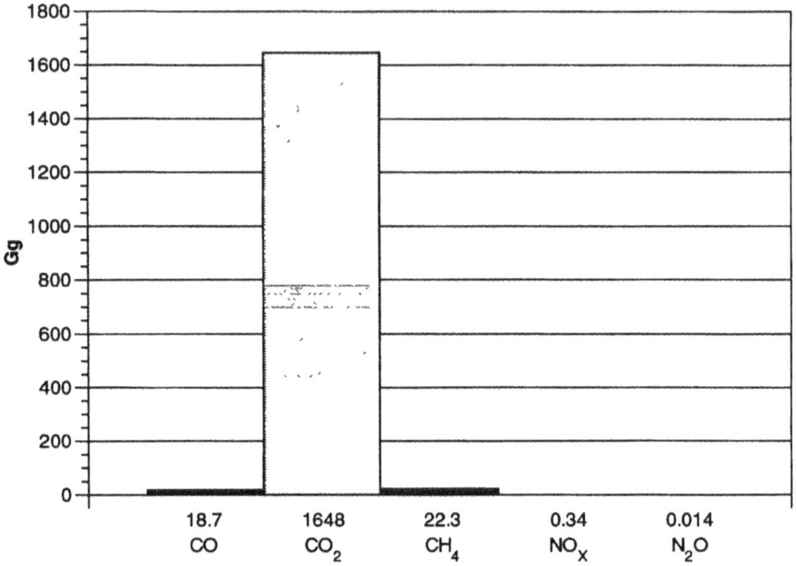

Figure 1. GHG emissions (Gg) by type of gas in the biomass sectors of The Gambia, 1993.

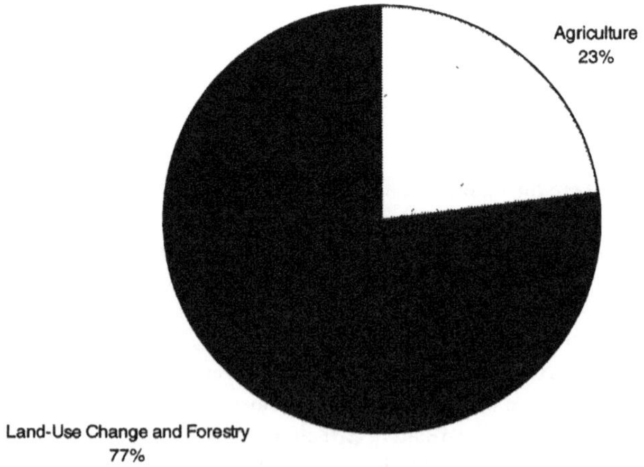

Figure 2. Share of emissions by biomass sectors of The Gambia in 1993.

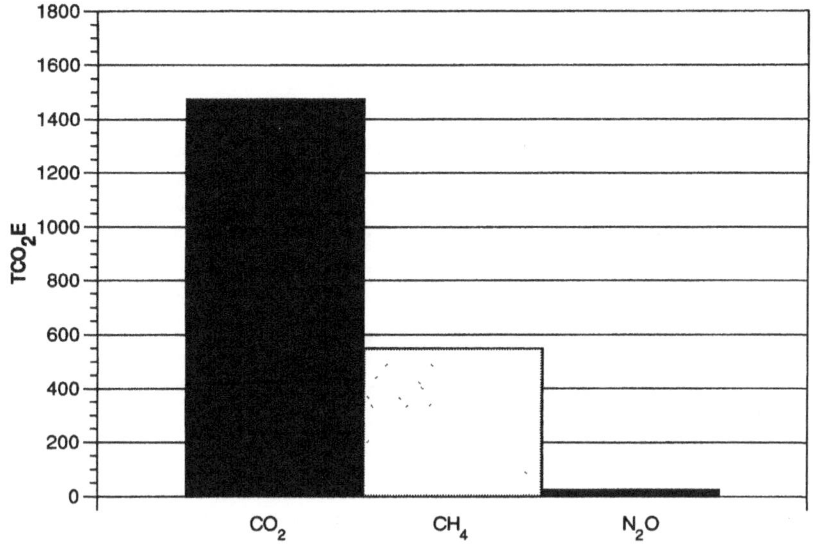

Figure 3. GHG emissions by gas from the biomass sectors in The Gambia (1993) in TCO_2E.

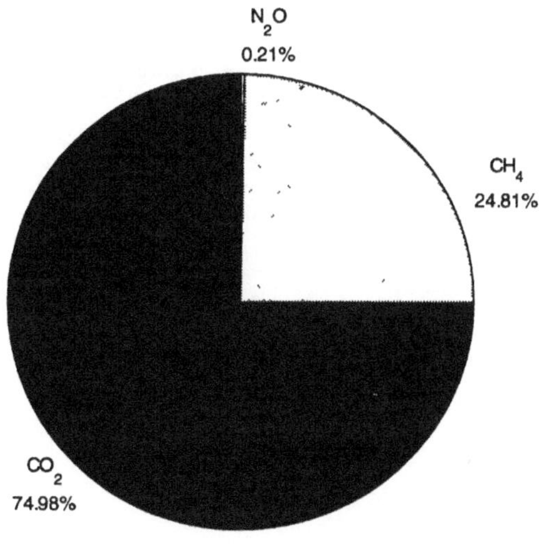

Figure 4. Share by gas of the total CO_2-equivalent emitted from the biomass sectors of The Gambia in 1993.

reported in this study, greenhouse gas emissions in The Gambia totaled 2 197 400 TCO_2E and, with a population of 1 025 867, the per capita emission is 2.142 TCO_2 per year.

4.2. Recommendations

In recognition of the contribution of The Gambia's biomass sectors to the greenhouse effect and its resultant global warming, it is recommended that The Gambia National Climate Committee develop programs for the mitigation of greenhouse gas emissions. These could be conducted nationally or in collaboration with other governments and institutions at the regional or global level, possibly through Joint Implementation.

5. References

Jabara, C.: 1990, Economic Reform and Poverty in The Gambia: A Survey of Pre- and Post-Economic Recovery Program Experience, Banjul, The Gambia.

Department of Forestry (DoF), The Gambia: 1991, Sectoral Contribution to The Gambia Environmental Action Plan (GEAP), Forestry Program.

Department of Livestock Services (DLS), The Gambia: 1992, Sectoral Consultation on Natural Resources and the Environment, Livestock Program.

Department of Planning (DoP), The Gambia: 1993, 1993 Statistical Year Book of The Gambia Agriculture.

Environment Unit, Ministry of Natural Resources and the Environment, The Gambia: 1992, The Gambia Environmental Action Plan.

Forster, H.: 1983, Evaluation of the National Forest Inventory of The Gambia, Technical Report no. 10, Gambia-German Forestry Project, Banjul.

Houghton, J.T., Callander, B.T., and Verney, S.K. (eds.): 1992, The Supplemental Report to the IPCC Scientific Assessment, IPCC, Cambridge, England: Cambridge University Press.

IPCC/OECD (Intergovernmental Panel on Climate Change/Organization for Economic Co- operation and Development) Joint Program: 1994, *IPCC Draft Guidelines for National Greenhouse Gas Inventories*, IPCC/OECD Joint Program, Paris, 3 Volumes.

Prince, S.D., Justice, C.O., and Los, S.O.: 1990, *Remote Sensing of the Sahelian Environment: A Review of the Current Status and Future Prospects,* Brussels: Centre Technique de Cooperation Agricole et Rurole (Commission of the European Community), Brussels, Belgium.

Senegal: 1994, 1991 Greenhouse Gas Emissions Inventory of the Republic of Senegal.

UNEP, OECD, IEA, IPCC (United Nations Environment Program, Organization for Economic Cooperation and Development, International Energy Agency, Intergovernmental Panel on Climate Change):1995, IPCC Guidelines for National Greenhouse Gas Inventories, IPCC, Bracknell, 3 Volumes.

NOTE

GREENHOUSE GAS EMISSIONS FROM FORESTRY, LAND-USE CHANGES, AND AGRICULTURE IN TANZANIA

R.S. MUYUNGI and C. OMUJUNI

Centre for Energy, Environment, Science, and Technology, P.O. Box 5511, Dar es Salaam, Tanzania

Abstract. The purpose of the study was to identify and quantify anthropogenic sources and sinks of greenhouse gases from forestry, land-use changes and agriculture in Tanzania. The 1990 inventory revealed that, in the land-use sector, methane (CH_4) and carbon dioxide (CO_2) are the primary gases emitted. Enteric fermentation in livestock production systems is the largest source of CH_4. Although deforestation results in greenhouse gas emissions, the managed forests of Tanzania are a major CO_2 sink.

1. Introduction

Tanzania, the largest country in East Africa, covers a total area of 88.6 million ha. The topography and climate are complex, with an altitude ranging from sea level to the summit of Mt. Kilimanjaro, the highest mountain in Africa (5895 m). The coastal areas are dominated by plains of up to 1000-m elevation dissected by river systems. The interior is characterized by plateaus at 1000 to 1500 m. The natural vegetation types include forests, woodlands, grasslands, bushland, and the coastal woodland. Woodlands and forests, estimated at 44 million ha, comprise about one-half the land area in Tanzania.

The economy of Tanzania is primarily agricultural. The agriculture sector employs about 84% of the working population and contributes about 61% of gross domestic product (GDP). Food-crop production dominates the agriculture sector, totaling 55% of agricultural GDP, with livestock accounting for 30%. The agricultural sector is based in 8000 villages, with an average farm of less than 2 ha per family.

Traditional cash crops include coffee, cotton, tobacco, tea, cashew nuts, and pyrethrum. Food crops include maize (*Zea mays* L.), rice (*Oryza* L.), wheat (*Triticum* L.), sorghum (*Sorghum* L.), millet (*Panicum* L.), cassava (*Manihot*), beans (*Vicia*), potatoes (*Solanum*), and bananas (*Musa* L.). There is a cattle herd of more than 13 million, 8 million goats, and 5 million sheep, all of which play an important role in the national economy. An estimated 90% of households keep livestock of some kind, primarily for subsistence. Cattle are particularly important for providing meat, milk, and manure. Within Africa, Tanzania ranks third in total livestock numbers after Sudan and Ethiopia.

The analysis of greenhouse gas (GHG) sources and sinks in Tanzania was a followup to an initiative of the Intergovernmental Panel on Climate Change (IPCC), under the auspices of the United Nations Environment Programme (UNEP) and the World Meteorological Organization (WMO) in response to growing concerns regarding global climate change (Houghton *et al.* 1992). To help the Government meet its obligations under

Environmental Monitoring and Assessment **38**: 313-316, 1995.

the United Nations Framework Convention on Climate Change (UNFCCC), the Tanzania Centre for Energy, Environment, Science and Technology (CEEST) was commissioned to characterize GHG sources and sinks in Tanzania. The main purpose of the study was to identify and quantify anthropogenic sources and sinks of GHGs in Tanzania.

2. Methods

Greenhouse gas emissions from land-use subsectors were analyzed using the UNEP/IPCC/OECD/IEA (1995) methods. The following GHGs were evaluated: methane (CH_4) emissions from rice cultivation; nitrous oxide (N_2O) emissions from the use of (N) nitrogen fertilizers; CH_4 emissions from enteric fermentation of domestic animals; CH_4 emissions from animal wastes; CH_4, carbon dioxide (CO_2), nitrogen oxide (NO_x), and N_2O emissions from the burning of agricultural crop wastes; and CO_2 sources and sinks of forest systems.

A simple methodology was used to estimate CH_4 emitted from rice fields, under which a daily emission rate range is used to compute annual emissions (Bachelet and Neue, 1993). In Tanzania, rice is cultivated under three regimes: flooded/irrigated, rain-fed, and upland. The number of days in a rice growing season is 120 days and 95 days for the flooded and rain-fed regimes, respectively. The average annual area of rice is 340.75 x10^3 ha (kha). Out of this, about 72 700 ha is upland rice area. The rice area under the irrigated regime is 41.9 x kha.

The emissions of CH_4 from enteric fermentation were based on various feeding systems for animals (Lerner *et al.*, 1988). Total CH_4 emissions depends on a number of factors, such as number and animal type, livestock management systems, size, and feeding and production characteristics. Simplified relationships for the estimation of CH_4 emissions were based on total animal feed intake and an estimate of the ratio of feed energy that is converted to CH_4.

Methane is generated from animal wastes when organic material decomposes anaerobically. The amount of CH_4 produced will depend on the type of waste, in particular the amount of volatile solids present in the waste. The amount of CH_4 emitted was calculated using the IPCC guidelines for national GHG inventories (UNEP *et al.*, 1995). The following assumptions were used in the calculations:

- Most cattle are grazed under the pastoral system, with the methane conversion factor (MCF) assumed at 0.01, at the maximum.
- In the case of poultry, the methane conversion factor of 0.01 was assumed.
- Swine, horses, goats, and sheep graze pastorally and an MCF of 0.01 was assumed.

Emissions of N are influenced by natural processes such as temperature, porosity, soil structure, precipitation, and farm management practices (for example, fertilizer type, crop type, and irrigation). Because interactions among the physical, chemical, and biological variables are complex, N_2O fluxes are variable in both time and space. Emissions of N_2O from fertilizer use were calculated using the *IPCC Guidelines for National Greenhouse Gas Inventories* (UNEP *et al.*, 1995).

The burning of crop wastes leads to the emission of CO_2, CH_4, CO, N_2O, and NO_x. Three crops have been considered in this analysis: cotton, sugarcane, and rice. Greenhouse gas emissions from the burning of these residues are estimated on the basis of the amount of C and N burned and the emission ratios of the gases (Ilukor and Oluka, 1995).

Land-use change results in a net exchange of GHGs. Because biomass is about 45% C by weight, clearing by burning leads to instantaneous release of CO_2, and also CH_4, CO, N_2O, and NO_x. When forests are converted to cultivated land, not all cleared aboveground biomass is burned in Tanzania. About 75% of the cleared biomass is collected as woodfuel, while approximately 20% is burned in the field, 10% of which remains on the ground as charcoal. The remaining 5% of aboveground biomass, such as foliage and twigs, decays in the fields over an average period of 10 years, releasing $^1/_{10}$th of its C content annually. Based on the quantity of CO_2 released from burning of aboveground biomass, GHGs are released instantaneously (Crutzen and Andreae, 1990).

Greenhouse gas emissions as a result of land-use change, especially deforestation, were estimated using methods and assumptions described by Dixon *et al.* (1993). Sequestration and storage of CO_2 by forest systems in Tanzania were estimated using the mean carbon (C) storage method. Carbon accretion and storage were calculated by summing the C standing crop for every year in the rotation and dividing by rotation length (Winjum *et al.*, 1992).

3. Results

The land-use sector of Tanzania produces significant GHG emissions (Table I). Within the agricultural subsector, enteric fermentation is the leading source of CH_4, followed by the burning of agricultural residues. Manure management, rice cultivation, and savanna burning are also significant sources of CH_4. The application of N fertilizer releases 0.57 Gg of N_2O. Combustion of savanna and agricultural residues produces insignificant amounts of N_2O, NO_x, and CO.

Forest systems are a significant CO_2 sink in Tanzania sequestering over 38 000 Gg annually. Vegetation regrowth on abandoned lands also sequesters CO_2. In contrast, conversion of forests and grasslands to agriculture releases small amounts of CO_2, CH_4, N_2O, NO_x, and CO. In balance, natural forest and agroecosystem of Tanzania sequester more CO_2 than released.

4. Discussion and conclusions

In 1990, the land-use sector of Tanzania was a net sink for CO_2, but a small source of CH_4, N_2O, NO_x, and CO from agricultural practices. Energy sector emissions in Tanzania are not significant given the agrarian socioeconomic conditions. Thus, Tanzania is not a major GHG producer in Africa (Houghton *et al.*, 1992).

TABLE I

Summary of greenhouse gas sources and sinks in the agriculture and land-use sectors in Tanzania in 1990 (Gg)

Source and sink	CO_2	CH_4	N_2O	NO_x	CO
Agriculture					
Enteric fermentation		845			
Manure management		26			
Rice cultivation		58			
Agricultural soils (fertilizers)			0.57		
Savanna burning		56	0.70	16	2 457
Burning of agricultural residues		327	0.57	13	1 053
Land-use changes					
Forestation and forest growth	−38 156				
Forest and grassland conversion	3 506	10	0.1	2	144
Reclamation of abandoned lands	−2 599				
Totals	−37 249	1 326	1.94	31	3 654

The *IPCC Guidelines for National Greenhouse Gas Inventories* (UNEP *et al.*, 1995) yield meaningful estimates of GHG sources and sinks in Tanzania. A systematic examination of transparent and common methodologies is needed within the land-use sector to narrow the variability of GHG emissions. A comparison of Tanzania GHG emissions to other countries reveals preliminary estimates are credible (Siddiqi, 1995).

References

Bachelet, D. and Neue, H.:1993, Methane emissions from wetland rice areas of Asia, *Chemosphere* **26,** 219–237.

Crutzen, P.J. and Andreae, M.O.: 1990, Biomass burning in the tropics: Impact on atmospheric chemistry and biogeochemical cycles, *Science* **250,** 1669–1678.

Dixon, R.K., Winjum, J.K., and Schroeder , P.E.: 1993, Conservation and sequestration of carbon: the potential of forest and agroforest management practices, *Global Environmental Change* **2,** 159–173.

Houghton, J.T., Callander, B.A., and Varney, S.K.: 1992, Climate Change 1992, The supplementary report to the IPCC Scientific Assessment, Cambridge, U.K.: Cambridge University Press.

Ilukor, J.O. and Oluka, S.O.: 1995, Carbon-to-nitrogen ratios in agricultural residues, *Environmental Monitoring and Assessment,* in press.

Lerner, J., Mathews, E., and Fung, I.: 1988, Methane emission from animals: a global high-resolution data base, Global Biogeochemical Cycles **2**, 139–156.

Siddiqi, T.: 1995, Asiawide emissions of greenhouse gases, *Annual Review of Energy and Environment* **20,** in press.

Winjum, J.K., Dixon, R.K., and Schroeder, P.E.: 1992, Estimating the global potential of forest and agroforest management practices to conserve and sequester carbon, *Water, Air, and Soil Pollution* **64,** 213–227.

UNEP, OECD, IEA, IPCC (United Nations Environment Programme, Organisation for Economic Co-operation and Development, International Energy Agency, Intergovernmental Panel on Climate Change): 1995, *IPCC Guidelines for National Greenhouse Gas Inventories,* IPCC, Bracknell, 3 Volumes.

LIST OF PARTICIPANTS

LIST OF PARTICIPANTS

Mr. Conde Abdoulaye
DCGTx/CCT, 04 BP 945
Abidjan 04, Côte d'Ivoire
Fax: 225-44 56 66
Tel: 225-44 53 56

Dr. Anthony O. Adegbulugbe
Centre for Energy Research and Development
Obafemi Awolowo University
Ile-Ife, Nigeria
Fax: 234-36-232975/
234-22410000
Tel: 234 36-233638/ 231740

Prof. Ayité-Lô N. Ajavon
Atmospheric Chemistry Laboratory
Université du Benin
Faculté des Sciences
BP 1515 Lomé, Togo
Fax: 228-218595/ 258595
Tel: 228-255094

Dr. Samir Amous
APEX-EBBD Consulting Co.
7 rue Malta Sghira, 1000 RP
Tunis, Tunisia
Fax: 216-1-835-583
Tel: 216-1-835-583

Ms. Susan Barvenik
ICF Incorporated
1850 K St., N.W., Suite 1000
Washington, D.C. 20006, USA
Fax: 202-862-1144
Tel: 202-862-1224

Ms. Barbara Braatz
ICF Incorporated
1850 K St., N.W., Suite 1000
Washington, D.C. 20006, USA
Fax: 202-862-1144
Tel: 202-862-1177
Email: braatzb@.icfkaiser.com

Dr. Bruce A. Callander
Intergovernmental Panel on Climate Change, Meteorological Office
Hadley Climate Centre, Room H202
London Road, Bracknell
RG12 2SY, United Kingdom
Fax: 44-1344-856912
Tel: 44-1344-856618
Email: bacallander.meteo.govt.uk

Dr. Renate Christ
Atmosphere Unit
UN Environment Programme
P.O. Box 30552
Nairobi, Kenya
Fax: 254-2-623410
Tel: 254-2-623447
Email: renate.christ@unep.no

Dr. Stuart Christie
Forestek, CSIR
Private Bag X11227,
Nelsprint 1200, South Africa
Fax: 27-1311-43869
Tel: 27-1311-43864

Environmental Monitoring and Assessment **38**: 319-325, 1995.

Mr. Andrew de Jode
Appropriate Technology
International, 67 East Ave.
Oxford, United Kingdom, OX4 1XR
Fax: 44-1-865-511-450
Tel: 44-1-865-248-119

Prof. Robert A. Delmas
Laboratoire d'Aerologie
U.R.A./C.N.R.S. 354
Université Paul Sabatier
118 route de Narbonne - F - 31062
Toulouse Cedex, France
Fax: 33-62-170988
Tel: 33-61-556467
Email: delr@aero.ups.tpse.fr

Dr. Robert Dixon
Director, U.S. Country Studies
Management Team
P.O. 63
1000 Independence Ave. SW
Washington, D.C. 20585, USA
Fax: 202-426-1540/ 1551
Tel: 202-426-0011
Email: csmt@igc.org.apc

Mr. Paul Drichi
National Biomass Study
Department of Forestry
P.O. Box 1613
Kampala, Uganda
Fax: 256-41-22-15-74
Tel: 256-41-25-17-78
Email: ghgug@mukla.gn.apc.org

Mr. William L. Driscoll
ICF Incorporated
9300 Lee Highway
Fairfax, VA 22201, USA
Fax: 703-934-9740
Tel: 703-934-3671

Dr. Yehia ElMahgary
VTT-Energy
P.O. Box 1606, 02044
VTT-Finland
Fax: 358-0-4566538
Tel: 358-0-4566563

Mr. Lemma Eshetu
Ethiopian Energy Authority
PO Box 486
Addis Ababa, Ethiopia
Fax: 251-1 51 78 74
Tel: 251-1 15 04 65

Dr. Aliou Faye
Program Officer, IUCN/Senegal
P.O. Box 3215
Dakar, Senegal
Fax: 221-32-92-46
Tel: 221-32-05-45

Mr. Jan Feenstra
Institute for Environmental Studies
Vrije Universiteit Amsterdam
De Boelelaan 1115
108 MV Amsterdam, Netherlands
Fax: 31-20-444-9553
Tel: 31-20-444-9555
Email: feenstra@sara.nl

Mr. Jack Fitzgerald
Project Manager, U.S. Country
Studies Management Team
P.O. 63
1000 Independence Ave. SW
Washington, D.C. 20585, USA
Fax: 202-426-1540
Tel: 202-426-0194
Email: jfitzgerald@igc.apc.org

Mr. R. Gacuhi
Ministry of Research, Technical
Training, and Technology
P.O. Box 30568
Nairobi, Kenya
Tel: 254-2-219-420

Dr. Greg Gaston
USEPA, Environmental Research
Laboratory
200 SW 35th St.
Corvallis, OR 97333, USA
Fax: 503-754-4799
Tel: 503-754-4496
Email: gregg@mail.cor.epa.gov

Mr. Michael J. Gibbs
ICF Incorporated
10 Universal City Plaza, #2400
Universal City, CA 91608, USA
Fax: 818-509-3137
Tel: 818-509-3186
Email: gibbsm@.icfkaiser.com

Mr. Asress Wolde Giorgis
Ethiopian Energy Authority
PO Box 486,
Addis Ababa, Ethiopia
Fax: 251-1 51 78 74
Tel: 251-1 15 04 65

Mrs. Sandy Guill
Program Manager for Africa
U.S. Country Studies Management
Team
P.O. 63
1000 Independence Ave., SW
Washington, D.C. 20585, USA
Fax: 202-426-1540/ 1551
Tel: 202-426-1464
Email: sguill@igc.apc.org

Ms. Ann Heidenreich
Climate Network Africa
P.O. Box 76406
Nairobi, Kenya
Fax: 254-2-72 9 447
Tel: 254-2-72 9 447
Email: aheidenreich@gn.apc.org

Ms. Rebecca Holmes
ICF Incorporated
1850 K St., N.W., Suite 1000
Washington, D.C. 20006, USA
Fax: 202-862-1144
Tel: 202-862-1176
Email: holmesr@mail.icfkaiser.com

Professor John Odeke Ilukor
Department of Physics
Makerere University
P.O. Box 7062
Kampala, Uganda
Fax: 256-41-251797
Tel: 256-41-531498
Email: jilukor@mukla.gn.apc.org

Dr. Augustine Isichei
Botany Department
Obafemi Awolowo University
Ile-Ife, Nigeria
Fax: 234-36-232-975/
234-36-233-784/
234-36-232-401
Tel: 234-36-230-290 x 3301

Mr. Bubu Jallow
Principal Meteorologist
Department of Water Resources
7 Marina Parade
Banjul, The Gambia
Fax: 220-228628
Tel: 220-228216

Dr. Vasantt K. Jogoo
VJ Environmental Services
76 Georgetown Complex
St. Jean Rd., Quatre Bornes
Mauritius
Fax: 230 454 4589
Tel: 230 454 4589

Mr. D. R. Kamdonyo
Meteorological Department
P.O. Box 2
Chileka, Malawi
Fax: 265 694 332
Tel: 265 694 355

Mr. Julius J. Kanyembo
Environmental Council of Zambia
PO Box 35131
Lusaka, Zambia
Fax: 260-1-223-123
Tel: 260-1-224-009/ 222678

Dr. G. Kasali
National Council for Scientific
Research
PO Box CH 158
Lusaka, Zambia
Fax: 260-1-223-123
Tel: 260-1-224-009

Mr. James M. Kasonde
National Council for Scientific
Research
PO Box CH 158
Lusaka, Zambia
Fax: 260-1-223-123
Tel: 260-1-224-009

Ms. Abeer S. Khalil
Organization for Energy
Conservation and Planning
32 Mahmoud Khairy St., Nasr City
Cairo, Egypt
Fax: 202-4037623
Tel: 202-607706/ 262-6349

Dr. Rik Leemans
Department of Terrestrial Ecology
and Global Change
RIVM, P.O. Box 1
3720 BA Bilthoven
The Netherlands
Fax: 31 30 292 897
Tel: 31 30 743 377
Email: rik.leemans@rivm.nl

Dr. Bo Lim
OECD Environment Directorate
2 rue Andre Pascal
75775 Paris Cedex 16, France
Fax: 33145247876
Tel: 33145247894
Email: bo.lim@oecd.org

Mr. James B. Magezi-Akiiki
Department of Meteorology
Ministry of Natural Resources
PO Box 7025
Kampala, Uganda
Fax: 256-41-251797/ 256166
Tel: 256-41-255609/ 251798
Email: ghgug@mukla.gn.apc.org

Dr. Mworia Maitima
National Museums of Kenya
P.O Box 40658
Nairobi, Kenya
Fax: 254-2-74-1424
Tel: 254-2-74-2131

Dr. Willy Makundi
Lawrence Berkeley Laboratories
MS90-4000, 1 Cyclotron Road
Berkeley, CA 94720, USA
Fax: 510-486-6996
Tel: 510-486-6852
Email: wrmakundi@lbl.gov

Dr. Caxton H. Matarira
Environment and Remote Sensing
Institute, P.Bag 7726
Causeway, Harare, Zimbabwe
Fax: 263-4-733-797
Tel: 263-4-731-045/9

Dr. Ruzvidzo S. Maya
Southern Centre for Energy and
Environment
31 Frank Johnson Ave.
Eastlea, Harare, Zimbabwe
Fax: 263-4-737351/ 739341
Tel: 263-4-737351
Email: scee@mango.zw

Mr. Abel Mbewe
National Energy Council
P.O. Box 37631
Lusaka 10101, Zambia
Fax: 260-1-252339/ 220301
Tel: 260-1-236017/ 236724

Mr. Dominic J. Mbewe
Department of Energy
P.O. Box 36079
Lusaka, Zambia
Fax: 260-1-252339/ 254491
Tel: 260-1-254686/ 251337

Mr. Kabelo K. Mogami
Department of Mines
Private Bag 0049
Gaborone, Bostwana
Fax: 09267 352141
Tel: 09267 352641

Mr. Mulalu Mulalu
Ministry of Agriculture
Private Bag 003
Gaborone, Botswana
Fax: 09267 356027
Tel: 09267 350500 (x 527)/ 350527

Mr. S. Abiodun Momodu
Centre for Energy Research and
Development
Obafemi Awolowo University
Ile-Ife, Nigeria
Fax: 234-36-232975/
234-2-2410000
Tel: 234 36-233638/ 231740

Mr. Harun R. Muturi
Ministry of Research, Technical
Training, and Technology
P.O. Box 30568
Nairobi, Kenya
Fax: 254-2-440-771
Tel: 254-2-219-420

Mr. Richard S. Muyungi
The Centre for Energy,
Environment, Science, and
Technology (CEEST)
1372 Karume Road, Oysterbay
PO Box 5511
Dar es Salaam, Tanzania
Fax: 255-51-66079
Tel: 255-51-67569/ 37256
Email: ceest@ceest.gn.apc.org

Ms. Cecilia Nzau
Ministry of Research, Technical Training, and Technology
P.O. Box 30568
Nairobi, Kenya
Fax: 254-2-440-771
Tel: 254-2-219-420

Dr. I. B. Obioh
Centre for Energy Research and Development
Obafemi Awolowo University
Ile-Ife, Nigeria
Fax: 234-36-232975/ 234-2-2410000
Tel: 234 36-233638/ 231740

Dr. Eric Odada
START/ENRICH Africa Liaison
Ghana Academy of Sciences
Agustino Neto Road, P.O. Box M32
Accra, Ghana
Fax: 233 21 777 655
Tel: 233 21 777 651-4

Mr. Seemadree A. Paupiah
Forestry Service
Botanical Gardens St.
Curepipe, Mauritius
Fax: 230-6743449
Tel: 230-6754966

Prof. Ademola Salau
UNDP/GEF, 1 UN Plaza
RBA-GEF
New York, NY 10017, USA
Fax: 212-906-5974/ 5423
Tel: 212-906-5911
Email: ademola.salau@undp.org

Dr. Bob Scholes
Forestek, CSIR, P.O. Box 395
Pretoria 0001, South Africa
Fax: 27-12-841-2689
Tel: 27-12-841-2045
Email: bscholes@forestek.csir.co.za

Dr. Faouzi Senhaji
GERERE
B.P. 6202 (R.I.)
10101 Rabat, Morocco
Fax: 212-7778135/ 674929/ 673550
Tel: 212-7777443
Email: faouzi.senhaji@gn.apc.org

Ms. Gillian Simmonds
Energy for Development Research Centre, University of Cape Town
Private Bag, Rondebosch, 7700
South Africa
Fax: 021-650-2830
Tel: 021-650-2419
Email: gill@energetic.uct.ac.za

Miss Iness Simuunza
Ministry of Environment and Natural Resources
P.O. Box 35131
Lusaka, Zambia
Fax: 260-1-223123
Tel: 260-1-224009

Dr. Youba Sokona
ENDA - Energy
PB 3370, 54 rue Carnot
Dakar, Senegal
Fax: 221-222695/ 235157
Tel: 221-225983/ 222496
Email: energy@endadak.gn.apc.org

Mr. Lloyd S. Thole
Environmental Council of Zambia
PO Box 35131
Lusaka, Zambia
Fax: 260-1-223-123
Tel: 260-1-224-009

Mr. Clive van Horen
Energy for Development Research
Centre, University of Cape Town
Private Bag, Rondebosch, 7700
South Africa
Fax: 021-650-2830
Tel: 021-650-2420
Email: clive@energetic.uct.ac.za

Mr. Graham P. von Maltitz
Forestek, CSIR
P.O. Box 395
Pretoria 0001, South Africa
Fax: 27-12-841-2689

Mr. Peter P. Zhou
EECG-Botswana
Box 402339
Gaborone, Botswana
Fax: 267-312 845
Tel: 267-312 845

GPSR Compliance

The European Union's (EU) General Product Safety Regulation (GPSR) is a set of rules that requires consumer products to be safe and our obligations to ensure this.

If you have any concerns about our products, you can contact us on ProductSafety@springernature.com

In case Publisher is established outside the EU, the EU authorized representative is:

Springer Nature Customer Service Center GmbH
Europaplatz 3
69115 Heidelberg, Germany

Batch number: 10761697

Printed by Printforce, the Netherlands